High Performance
Data Acquisition and Control

J.W. Tippie, Ph.D.
D.A. Cooper, Ph.D.
R.T. Cleary

High Performance Data Acquisition and Control, First Edition

Copyright © 1994 by KineticSystems Corporation
All rights reserved. No part of this book may be reproduced,
in any form or by any means, without permission of
KineticSystems Corporation.

Publisher's Cataloging-in-Publication Data

Tippie, J.W.
    High performance data acquisition and control/
    J.W. Tippie, D. A. Cooper, R.T. Cleary, et. al.—
    Lockport, IL: KineticSystems Corp., [1994]
       p.   ill.   cm.
    Includes bibliographical references and index.
      ISBN: 0-9640525-0-4
      1. Automatic data collection systems. I. Cooper, D.A.,
    II. Cleary, R.T., III. KineticSystems Corporation. IV. Title.
    TS158.6.T57   1994
006-dc20

Cover Design by Knorr Marketing

Printed in United States of America

10 9 8 7 6 5 4 3 2

Second Printing, 1994

About KineticSystems Corporation

KineticSystems was founded in 1970 in Lockport, Illinois. For more than two decades, the company's business focus has been the development of hardware and software products and systems for computer-controlled, high-performance data acquisition and control.

The company's hardware products are based on internationally-accepted standards: CAMAC, VME, and VXI. Its data-acquisition systems are compatible with MS-DOS, UNIX, POSIX and OSF standards.

KineticSystems maintains an ongoing commitment to meet the data acquisition needs of science and industry. The company's customer base is worldwide and includes all major scientific laboratories and one hundred fifty colleges and universities. It also includes most major aerospace and defense contractors, the major automotive manufacturers and about fifty other Fortune 500 companies.

KineticSystems is a Cooperative Marketing Partner with Digital Equipment Corporation and a Preferred Solution Provider with the Computer Division of Hewlett-Packard.

About the Authors

Dr. J.W. Tippie *is a graduate of Massachusetts Institute of Technology and received his Ph.D. from Case Institute of Technology in Physics. He is the author of many papers in both physics and the computer-based data acquisition field. He has over 30 years experience in the data acquisition and control field, including 15 years at a major scientific laboratory designing and implementing data acquisition systems. He was responsible for the conceptual design of the Tachion System which was awarded the Industrial Research IR 100 award, as well as the primary architect of the KineticSystems Silver Bullet VXI product line that was awarded the VXI Product of the Year award by the VXI Journal.*

Dr. David A. Cooper *has over 20 years experience designing data acquisition systems for scientific and industrial applications. He received his B.Sc. from the University of Leicester, England and his Ph.D. from the University of Durham, England. Dr. Cooper has published over 20 technical papers and has taught both undergraduate and graduate level courses in Electronics at the University of Utah.*

Mr. Robert T. Cleary *has been involved with the technical aspects of communications and data acquisition for over 30 years. After receiving a BSEE degree in Electrical Engineering from the Illinois Institute of Technology in 1960, he joined GTE Automatic Electric Laboratories as a development engineer. While at GTE, he developed many new products and was issued 18 U.S. patents and 28 foreign patents. In 1970 he was co-founder of KineticSystems and is currently its Chief Executive Officer. Mr. Cleary is responsible for developing many of the techniques used for high-performance data acquisition.*

Contents

1 Introduction   1
   1.1 High-Performance Data Acquisition 1
   1.2 Data Quality 2

2 Standards   5
   2.1 Hardware Standards 5
       2.1.1 Backplane Buses 6
       2.1.2 Interconnects 6
       2.1.3 Networking 6
       2.1.4 Advantages 6
   2.2 Software Standards 6
       2.2.1 POSIX 7

3 Sampling Data   9
   3.1 Sampling Theorem 9
   3.2 Aliasing 10
       3.2.1 Antialiasing Filtering 11
       3.2.2 Establishing a Sampling Rate 12

4 Analog Front-End   13
   4.1 Proper Connections to the Sensors 13
       4.1.1 Differential vs. Single-ended Input Channels 13
       4.1.2 Common Mode and Normal Mode 15
       4.1.3 Isolation 16
       4.1.4 Grounding—Bad Ground Loops vs. Good Ground Loops 17
   4.2 Noise 23
   4.3 Calibration 24
   4.4 Gain 24

	4.5	Auto-Ranging	25
	4.6	Signal Conditioning	25
		4.6.1 Bridge Conditioning	25
		4.6.2 Thermocouples	26
	4.7	Reading the Analog Specifications	27
		4.7.1 Inputs	27
		4.7.2 Gain Ranges	29
		4.7.3 Frequency Response	29
		4.7.4 Bridge Completion	30
		4.7.5 Shunt Calibration	30
		4.7.6 Accuracy	30
		4.7.7 Gain Stability and Offset Voltage Stability	30
		4.7.8 Noise	31
		4.7.9 Linearity	31
		4.7.10 Common Mode Rejection Ratio	31
		4.7.11 Bridge Balance	32
5	**Filtering**		**33**
	5.1	Antialias Filter Considerations	34
		5.1.1 System Costs	34
		5.1.2 Filter Costs	35
	5.2	Types of Filters	35
		5.2.1 Bessel Filters	35
		5.2.2 Butterworth Filters	36
		5.2.3 Chebyshev Filters	36
		5.2.4 Elliptic or Cauer Filters	36
	5.3	Filter Implementations	36
		5.3.1 Continuous Filters	36
		5.3.2 Switched Capacitor Filters	37
	5.4	Choosing Filters and Sampling Rates	37
		5.4.1 ADC Quantization Noise	38
		5.4.2 Signal Characteristics	38
		5.4.3 Selecting the Filter and the Sampling Rate	39
	5.5	Digital Filtering	40
6	**Analog-to-Digital Converters**		**43**
	6.1	Types of ADCs	43
		6.1.1 Integrating ADC	43
		6.1.2 Successive Approximation ADC	43

	6.1.3	Flash Converters	44
	6.1.4	Hybrid Converters	44
	6.1.5	Sigma-Delta Converters	44
6.2	Resolution and Accuracy		45
6.3	Multiplexing Analog Signals		45
6.4	Sequential Scan vs. Simultaneous Sample-and-Hold		46
	6.4.1	Simultaneous Sampling	46

7 Digital Processing and Buffering — 47

7.1	Throughput	47
7.2	Latency	48
7.3	Operating System Latency	49
7.4	Buses, Bandwidth and Latency	50
	7.4.1 Backplane Buses	51
	7.4.2 Interconnect Buses	52
7.5	Buffering and How It Helps Overcome Latency	53
	7.5.1 Buffering Techniques	54
7.6	Using Dedicated Processors to Overcome Latency	56

8 Types of Data Acquisition — 59

8.1	Continuous	59
8.2	Transient	60

9 Systems Considerations — 63

9.1	Calibration	63
	9.1.1 Voltage Injection	64
	9.1.2 Bridge Transducer Verification	64
	9.1.3 Reference Voltage	64
	9.1.4 Calibration Archiving	65

10 Software Considerations — 67

10.1	Data Acquisition Software Features	67
	10.1.1 Large versus Small System Considerations	69

11 System Considerations and VXI — 71

11.1	VXI Bus Features	71
	11.1.1 VXI Local Bus for Inter-module Communications	71
	11.1.2 Local Bus for Routing Analog Signals	72
	11.1.3 Local Bus as a Private Digital Path	74

	11.1.4 Data Flow	75
11.2	Trigger Lines for Time Synchronization	75
11.3	Overall Architecture	76
11.4	System Calibration and Diagnostics	78
11.5	System Configuration Validation	79

A VXIbus, an Open Instrumentation Standard   81

A.1	VXI Module Sizes	82
A.2	The VXI Mainframe	82
A.3	The Slot-0 Controller	83
A.4	Using the Module ID Lines	84
A.5	Register-Based vs. Message-Based Modules	85

Bibliography   89

Index   91

List of Tables

2.1	The POSIX Family of Standards	7
4.1	Typical Specifications for Bridge Completion Signal Conditioning	28
5.1	Theoretical Signal to Noise Ratios for Typical ADCs	39

List of Figures

3.1	Signal aliasing illustration	11
4.1	A single-ended data acquisition input channel	14
4.2	A differential data acquisition input channel	15
4.3	Measuring common mode rejection ratio	16
4.4	An isolated data acquisition input circuit	17
4.5	Various shield grounding methods	19
4.6	A data acquisition system at 1 million volts above ground	20
4.7	Driving a differential input channel from an unbalanced source	21
4.8	Connecting drivers with coaxial outputs to differential inputs	22
4.9	Typical thermocouple configuration	27
5.1	An 8-Pole Butterworth filter illustrating sampling rate set so aliased signals above accuracy threshold fall back to f_c for frequency domain analysis applications.	41
7.1	Typical OS Process latency histogram	50
11.1	Typical signal conditioning module	72
11.2	Typical MUX-bus configuration	73
11.3	Typical Digi-bus configuration	75
11.4	Typical system configuration illustrating MUX-bus and Digi-bus	76
11.5	Typical ADC configuration	77
11.6	Typical calibration configuration	79
A.1	Relative sizes of VXI module options	83
A.2	VXI host computer options	84
A.3	Message-based and register-based VXI options	86

Chapter 1

Introduction

There are many aspects to data acquisition which range from the sensors ...to the analog signals that represent some measured quantity ...to the sampling and digitization of the signal ...to buffering the data into a computer ...to archiving it to some storage device ...to analyzing the data and extracting some information about the process being monitored. In the final analysis, what is important is that the data acquisition system capture the relevant data with sufficient accuracy to permit the reliable extraction of the required information. Obviously this process starts with the design of the measurement environment and its sensors, and ends with some information about the process that is being measured. Each step of this process is critical, for once *information* is lost, it can never be recovered.

This book examines the practical issues of acquiring accurate and reliable data, particularly as they relate to high-performance applications. Various tradeoffs affecting cost and performance as well as some of the "systems-level" issues that make effective data acquisition systems are discussed. The book traces the data acquisition process from the sensor output through archival of the data for analysis. While the process of selecting appropriate sensors and choosing the critical points to measure is extremely important, these considerations are beyond the scope of this book. Also, data analysis techniques, except as they may relate to the data acquisition process, are not addressed.

1.1 High-Performance Data Acquisition

High-performance data acquisition relates not only to speed and throughput, but also involves matters of data quality, accuracy, and flexibility to adapt to varying situations.

1.2 Data Quality

The subject of *data quality* embraces a number of concerns. The ultimate goal is to capture the critical information. Data and information are not necessarily the same. It is easy to collect a lot of meaningless data, but not always as easy to capture the required information.

Many factors can influence the quality of the data collected. These include the accuracy of the transducers, amplifiers, multiplexors, and ADCs; noise contributed by each of the system components; pickup of stray electro magnetic fields; and distortion introduced by non-linear and frequency-dependent components. Also included are errors introduced by high frequency signals aliased into the passband by the sampling process. The accuracy of calibration of system components and their traceability to NIST standards can also affect data quality.

In specifying a data acquisition system, understanding the degree of data accuracy that is required to extract the necessary information, and designing the system with a consistent set of requirements are important. For example, the system is not matched to the requirements if one specifies an ADC accuracy to 16-bits (0.0018% accuracy) for an application where the transducer is only accurate to 0.1 percent, or fails to specify sufficient high-frequency noise rejection in a high EMI environment. In the first case, the system price becomes excessive for the requirement, and in the second, the data quality is lost.

The key to specifying a cost-effective solution is a balance between the cost and the performance specifications of each component. As performance specifications of components are held to tighter requirements, costs increase rapidly as one approaches the leading edge of the state-of-the-art. Over-specification can easily double or triple system costs. In general, sources of error contributing to the over-all system performance are statistically independent and add as square root of the sum of the squares of the individual errors (σ_i).

$$\sigma_{\text{system}} = \sqrt{\sum_i \sigma_i^2}$$

Thus, in our example, moving from an ADC accurate to 12 bits (0.03%) to an ADC accurate to 16 bits (0.0018%) has negligible effect on the overall accuracy of the system. In fact, the ADC with 12-bit accuracy is more than sufficient in this case.

$$\sqrt{(0.1)^2 + (0.03)^2} = \sqrt{0.01 + 0.0009}$$

versus

1.2. DATA QUALITY

$$\sqrt{(0.1)^2 - (0.0018)^2} = \sqrt{0.01 + 0.00000324}$$

The above discussion indicates that any source of error under about 30% of the dominant error will have negligible effect (less than 5%) on the overall system error. When establishing the accuracy specifications for a data acquisition system it is also important to consider the long term accuracy needs, including those that may arise from improved sensor technology.

Chapter 2

Standards

Standards are playing an increasing role in the selection of data acquisition front-ends as well as computer hardware and software. Standards allow system designers to choose from a wider range of equipment options; provide flexibility by making it possible to upgrade systems to meet unanticipated needs; make it possible to integrate parts of one system that was designed to solve one problem into a system that solves another problem; permit one system to communicate with another; allow applications to be ported from one platform to another; and make it easier to upgrade those parts of systems where technology is evolving rapidly. The rapid evolution of computer CPUs is a good example of this.

Standards in the data acquisition and computer field take two forms: formal standards developed and sponsored by such non-profit organizations as IEEE, ANSI, and ISO; and *defacto industry standards* such as the PC. In some cases where "industry standards" become widely accepted, they are adopted by standards organizations. HPIB, for example, was developed by Hewlett Packard and was adopted with some minor changes for GPIB (IEEE-488).

2.1 Hardware Standards

Hardware standards in the data acquisition and computer field have a long history going back to the NIM, CAMAC (IEEE 583), GPIB (IEEE 488), and ANSI RS-232 standards. More recent standards include VME, VXI, SCSI, FDDI, Futurebus PLUS and Fastbus.

Important standards in data acquisition include backplane buses such as VME and VXI, device interconnects such as SCSI and GPIB, and networks such as Ethernet and FDDI. These standards provide the basis for interconnecting the various components of a data acquisition system.

2.1.1 Backplane Buses

Backplane buses provide a powered, high-performance bus, that allows data acquisition components to be tightly integrated with computer system components. For example, a computer system that provides a VME bus can be used with VME- or VXI-based data acquisition boards.

2.1.2 Interconnects

Device interconnects, such as SCSI and GPIB, are frequently available on various computer platforms and can provide a high-performance connection to various instruments and data acquisition devices. These interconnects are typically used to connect system components in separate chassis or boxes.

2.1.3 Networking

Networking has become increasingly important and has allowed data to be acquired at one location and analyzed in one or more remote locations. Low-cost processors and workstations have made an important impact in this area. They have also made it possible to put the computing power closer to the point of acquisition—particularly in control applications where the control loop can now be placed in the same chassis or module as the I/O. This technique minimizes the need to provide high-performance data paths back to some remote processor and the need for low-latency response by large, general purpose systems.

2.1.4 Advantages

In general, hardware standards allow users to select components based on performance requirements and integrate the pieces into a complete application. Standards provide the basis that enables users to interconnect various subsystem components. Standards also facilitate the long-term evolution of a system. Many applications evolve over time. By using standards, it is *much* easier to adapt a system to meet requirements.

2.2 Software Standards

Software standards are becoming increasingly important. POSIX (IEEE 1003) as well as ANSI C and Fortran are available today. Many defacto standards also exist, e.g. MS-DOS, TCP/IP, and X-Windows.

2.2. SOFTWARE STANDARDS

Table 2.1: The POSIX Family of Standards

POSIX.0 is a guide to POSIX Open Systems Environment.

POSIX.1 defines the interface between portable applications and the operating system and includes C language bindings.

POSIX.2 defines the shell command language.

POSIX.3 provides verification and testing requirements for the POSIX family.

POSIX.4 defines a set of realtime extensions to POSIX.1.

POSIX.5 defines the ADA bindings to POSIX.1 services.

POSIX.6 defines the security enhancements.

POSIX.7 lists the standards for system administration.

POSIX.8 defines the standards for distributed file systems, protocol independent network interfaces, remote procedure calls, and interfaces to seven-layer OSI protocol.

POSIX.9 defines the FORTRAN language bindings to POSIX.1 services.

POSIX.10 defines the Supercomputing Application Environment Profile (AEP).

POSIX.11 defines the Transaction Processing AEP.

2.2.1 POSIX

The POSIX or *Portable Operating System Interface for Computing Environments* specification (IEEE 1003.1) defines a standard for an application to interface with an operating system. POSIX defines a set of functions based on a combination of AT&T UNIX System V and Berkeley Standard Distribution UNIX. It is important to note that POSIX defines the interface to the operating system(OS)—not the underlying operating system internals. The intent is to provide a basis for portable applications code.

POSIX is actually a series of proposed standards with 1003.1 defining the basic operating system interface[4]. Table 2.1 gives a brief summary of the proposed standards.

Chapter 3

Sampling Data

Most data acquisition systems involve the sampling of one or more continuous analog signals. This time-series of discrete samples is typically used to represent the original continuous time-varying signal. To accurately represent a continuous time-varying signal, it is necessary to sample the signal at a sufficiently high rate to correctly represent the time dependance of the signal.

Also, when this discrete time-series of samples is used to extract time- or frequency-dependent information present in the original signal (typically the case), a phenomenon known as *aliasing* can result. This can cause a significant loss of accuracy and lead to misconceptions about the process being measured. Note that aliasing *does not* infer that the "sampled value" is in error, but rather means a distortion of the inferred time dependence of a series of samples. Refer to Figure 3.1.

Aliasing occurs whenever there are frequency components in the original signal above the Nyquist frequency—half the sampling rate—at the point where the signal is sampled (typically at the ADC). A common oversight is to inject a nice "clean" sine wave from a generator into a data acquisition system, then under-sample the signal. When the resulting digitized waveform is displayed, it appears to have little resemblance to the *expected* sine wave, causing confusion about whether the signal generator or the data acquisition system has possibly malfunctioned.

3.1 Sampling Theorem

The ultimate goal of any data acquisition system is to capture critical *information* about the process being monitored. Since digital data acquisition by nature must sample the various signals present, the question is, "How fast should the sampling be done to acquire the desired *information*?" Obviously, the process and sensor play crucial roles. Generally, the dynamics of the process and the sensor, limit the generated signal bandwidth. For example, a thermocouple buried in a copper block will only respond so fast when thrust into boiling water.

Furthermore, the information of interest may lie at a much lower frequency compared to the process dynamics and possibly the sensor. For example, a simple undamped float for measuring the gasoline level in a car would be almost useless, since the information needed is the average level of fuel, not how much it "sloshes" around when the car is moving.

The answer to the sampling question is that a data acquisition system *must* sample at a rate of at least twice the highest frequency of interest. It *must* also limit the bandwidth of the signal *prior* to sampling to avoid aliasing (see Section 3.2).

In the thermocouple case, the thermal time constant typically limits the maximum rate of change of any signal observed and thus the maximum frequency of interest. The minimum sampling rate is twice the maximum frequency of interest.

The undamped gasoline gauge is more complex. Here, the *maximum frequency of interest* is well below the sensor signal bandwidth—probably an effective time constant of several minutes works. It is necessary to *average* the input signal in some way to eliminate the *high frequency noise* due to the fuel sloshing in the tank. This can be accomplished in several ways. One technique is to use a low-pass filter before the signal is sampled. The filtered signal is then sampled at a minimum rate of twice the highest frequency in the filtered signal. The alternative is to sample at a much higher rate determined by the raw sensor signal bandwidth and *average or filter* the signal using digital techniques.

Even in the thermocouple case, filtering is often desirable to attenuate noise introduced on the sensor cable.

3.2 Aliasing

Signal aliasing occurs whenever a signal is sampled that has frequency components above the Nyquist frequency (half the sampling frequency). A simple illustration of this phenomenon is shown in Figure 3.1. In this figure, a 900 Hz sine wave is sampled at 1000 Hz. The 900 Hz signal is illustrated in light gray and the sampled points by circles. If one looks only at the sampled points, one *infers* that the original signal was a 100 Hz sine wave, i.e., the 900 Hz signal was *aliased* to 100 Hz. Obviously, had one sampled the 900 Hz signal at a rate above 1800 Hz (twice the highest input frequency), the inferred signal would be 900 Hz.

Note that the aliasing effect has to do with the act of "sampling" and applies whether the "signal" is a continuous analog signal or a series of discrete digital values. For example, "sampling" every 10th point from a series of digitized points representing a time

3.2. ALIASING

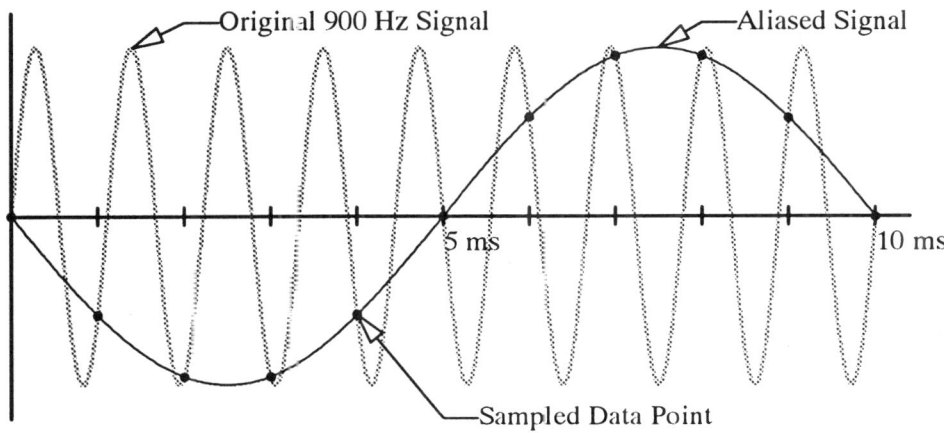

Figure 3.1: Signal aliasing illustration

dependent signal can result in just as much of an aliasing effect as having digitized the original signal at 10th the sampling rate. The graph in Figure 3.1 is, in fact, composed of discrete samples at 300 dots per inch. Also note that once a signal is aliased, there is no magic to undo the aliasing.

To take this one step further, if one could imagine a frequency domain plot of a signal, the effect of aliasing is to fold the plot accordion-style at multiples of the Nyquist frequency. Thus, a frequency component just above the Nyquist frequency aliases an equal distance below it. A frequency just below twice the Nyquist aliases just above zero frequency, etc.

3.2.1 Antialiasing Filtering

The question frequently is "Does one need to worry about aliasing?" The answer is probably *yes* unless the transducers are low bandwidth and one can afford to over-sample, or one is lucky enough to have one of those few applications where all the information is in the sampled signal value, like the charge deposited in a detector by a nuclear particle, and *not* in the time history of the signal.

The other solution to the aliasing question is to limit the bandwidth of the signal *before* it is sampled. This is accomplished with a low pass filter prior to any multiplexing or sampling *(always remember that aliasing is a sampling phenomenon)*. The low-pass filter limits the bandwidth of the signal prior to sampling. Such a filter is required for the hypothetical gasoline gauge discussed in Section 3.1. A more in-depth discussion on filtering is found in Chapter 5.

3.2.2 Establishing a Sampling Rate

One very serious decision which the designer of a data acquisition system must make is that of sampling rate. No one wants to miss information or get erroneous results from aliasing. Therefore, one might conclude that the simplest answer is a "high" sampling rate. The down side of this solution is typically cost, particularly where there are many I/O points and/or a high bandwidth requirement. Most modest systems will fairly easily accommodate an aggregate throughput (sampling rate of each channel times number of channels) of 1000 samples/second. Aggregate throughputs to 100,000 samples/second are not too difficult to achieve with today's technology. Generally system costs rise quickly for aggregate sampling rates above 500,000 samples/second.

As a general rule, a little effort to limit the amount of data is well placed since it reduces system cost, storage requirements, and analysis time. This is especially true when dealing with higher throughput applications.

Chapter 4

Analog Front-End

Most data acquisition systems involve obtaining data from various transducers that produce analog signals. Often, the signals from the transducers are low level and require various kinds of signal conditioning. Also, the transducers are frequently located some distance from the data acquisition front-end.

This chapter explores some of the issues involved with field wiring, grounding, isolation, signal conditioning, noise, calibration and related *analog* issues.

4.1 Proper Connections to the Sensors

Measuring Data and Not Noise

4.1.1 Differential vs. Single-ended Input Channels

Two very important concepts that affect the performance of a data acquisition system are *single-ended* and *differential*. A single-ended input channel, as shown in Figure 4.1(a), completes the circuit from a sensor to the data acquisition (DAQ) system input circuit via a *signal* wire and a *return* wire. The *return* wire is usually the cable shield and is generally connected to the DAQ system's circuit common—which is generally connected to "ground." In an ideal world, this should not present a problem. *Unfortunately, the world is filled with many noise sources that can interfere with data.* The problem with single-ended input circuits is that the cable shield is part of the signal path—*and any noise voltage developed across the shield adds full-force to the signal!*

Figure 4.1(b) shows a sensor that is connected to "ground" and is wired to a grounded

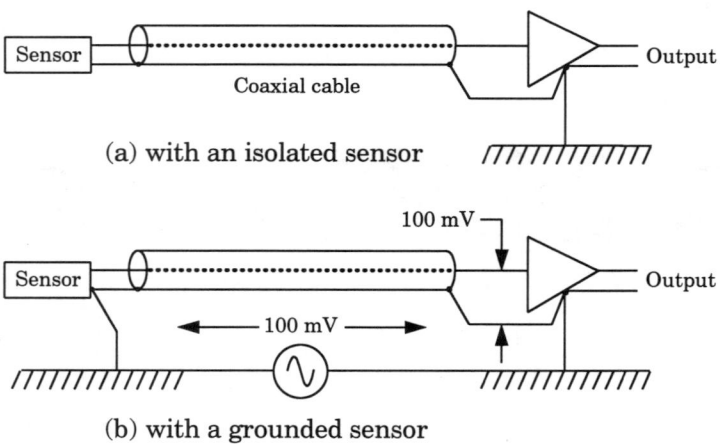

(a) with an isolated sensor

(b) with a grounded sensor

Figure 4.1: A single-ended data acquisition input channel

single-ended input. If there is 100 millivolt ac potential difference between the sensor's "ground" and the DAQ system's "ground," then *all of the 100 millivolt noise* will be superimposed on the signal. If the sensor is a thermocouple, the noise is likely larger than the dc signal voltage. Even if the sensor has no connections to ground, the data acquisition system must be carefully designed so all connection points are very close to the same potential, or errors will be introduced to the received signal of the various channels. Also, if the input wiring is close to sources of electrical noise, interference may be coupled into the signal path.

A differential-input channel—often called a *balanced input*—is generally connected to a sensor as shown in Figure 4.2(a). A *shielded twisted pair* cable is most often used for differential operation. The signal is received by an *instrumentation amplifier*. The primary characteristic of an instrumentation amplifier is that it delivers a signal at its output that is proportional to the *difference* between the voltage on its "+" input and its "-" input.

A real-world example of a differential-input channel is shown in Figure 4.2(b). Note that, in this case, the signal circuit is completed *without any signal passing through the shield*. If a noise voltage is impressed across the shield because of a ground potential difference, the effect on the signal will be greatly attenuated. The characteristic of the cable can cause interference to creep into a differential system with long cable runs. The shielded cable must contain a *twisted pair*. Some single-pair cables have both conductors

4.1. PROPER CONNECTIONS TO THE SENSORS

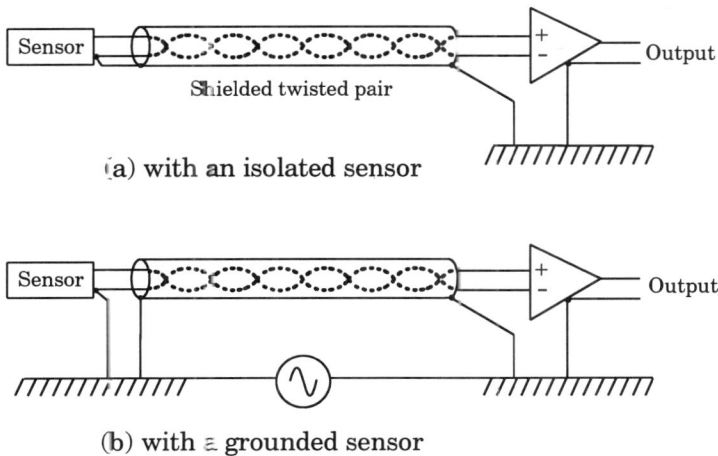

(a) with an isolated sensor

(b) with a grounded sensor

Figure 4.2: A differential data acquisition input channel

in the same plane instead of having the pair of wires twisted. If this cable is near a strong source of electrical noise, one signal wire will be nearer to the noise than the other, and the noise will not cancel as well. Also, some two-pair cables with separate shields do not contain pairs that are twisted. Systems using such cables can exhibit unacceptable coupling—called crosstalk—between the channels, even though each pair is shielded.

4.1.2 Common Mode and Normal Mode

Common mode rejection ratio—often abbreviated CMRR—describes the effect that unwanted noise between the signal conductors and ground has on the desired input signal. It is called *common* mode because the unwanted signal (often caused by power line noise) is impressed across both conductors of a differential pair and—in the ideal case—is cancelled out by the balanced system. CMRR is generally measured as shown in Figure 4.3. The test voltage is impressed across both conductors of the differential input. Often this test uses a 1000-ohm resistor in one leg to represent the fact that the "real" transducer source may not be perfectly balanced. CMRR is usually expressed in decibels (dB). In this case the dB value represents a logarithmic voltage ratio between the output signal caused by a common-mode voltage and that from a normal-mode voltage. A normal-mode voltage is that impressed *across* the input conductors. For single-ended systems with the shield conductor connected to ground at the instrumentation front-end, the ratio between

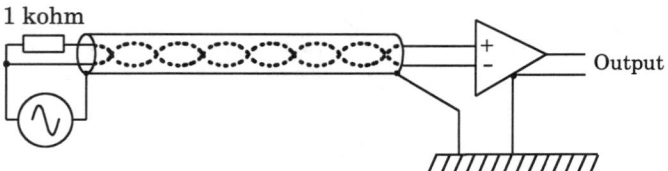

Figure 4.3: Measuring Common Mode Rejection Ratio

common mode and normal mode is 1:1—*no common-mode rejection*.

Each 20 dB represents an increase in the common mode rejection ratio by a factor of 10. Therefore a CMRR of 80 dB represents an attenuation of common-mode noise equal to 10,000 to 1. Another important parameter here is the maximum linear input swing on the input circuit. This value is often ±10 volts. With a CMRR of 80 dB, a 10 volt RMS (Root Mean Square) common-mode signal should have an effect on the input signal of 2 millivolts (20/10,000). However, if this is a sine wave, the voltage will reach positive and negative peaks of about 14 volts. This will likely be outside the linear range of the instrumentation amplifier and will result in substantial feed-through during portions of each cycle.

As just discussed, the value of CMRR determines how much *common-mode* noise gets converted to *normal-mode* signal. Some amount of noise may also *enter the system as normal-mode signal*, caused by noise pickup at the transducer, etc. For whatever cause, once an unwanted signal becomes a normal-mode voltage, it can be eliminated only by filtering. This is usually accomplished by low-pass filtering. For slowly changing signals, such as thermocouples, this is often accomplished by one- or two-pole passive (resistor-capacitor) filters connected directly to the input pair before any electronics. The cutoff frequency for these filters is often in the range of 2 to 10 Hertz to provide good normal-mode attenuation to power line frequency and its harmonics.

For fast-changing signals, any attempt to attenuate 50 or 60 Hz power line frequencies would prevent these changes from being monitored by the data acquisition system. In this case, sufficient precautions—such as proper grounding and good CMRR—must be taken to prevent noise from becoming normal-mode in the first place.

4.1.3 Isolation

As indicated earlier, most instrumentation front-ends require that each conductor of a differential input—and the signal conductor of a single-ended input—remain within about

4.1. PROPER CONNECTIONS TO THE SENSORS

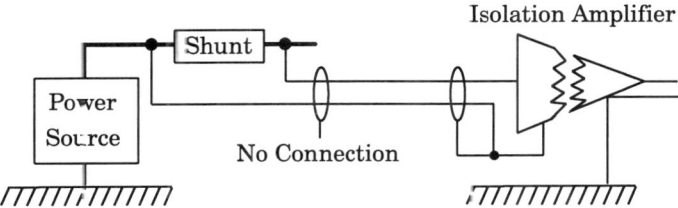

Figure 4.4: An isolated data acquisition input circuit

±10 volts of the input common ground. This is not satisfactory for input signals with large common-mode voltage present. An example of this is the measurement from a current shunt that is several hundred volts above ground. An isolated input circuit is appropriate for such an application. This isolation can be built in to the data acquisition system or consist of an isolation block in front of the DAQ system. In either case, the most common practice is to use an isolation amplifier for this purpose. An isolation amplifier usually uses a dc-to-dc-converter-powered "floating" amplifier that produces a high-frequency signal whose duty cycle is proportional to the input voltage. This high-frequency signal is then coupled across an isolation barrier and filtered. The resulting signal is proportional to the input voltage and is coupled via the output amplifier. An isolated input circuit monitoring a "floating" shunt is shown in Figure 4.4. The advantage of such an input circuit is that it can accommodate high common-mode voltages. Isolated input channels result in a significant increase in per-channel cost. Frequency response is usually limited to about 30 kilohertz. Isolated input circuits are usually specified only when a non-isolated approach cannot be used.

4.1.4 Grounding—Bad Ground Loops vs. Good Ground Loops

The subject of good grounding practices usually causes more arguments than any other aspects of high-performance data acquisition. Also, it is generally felt that ground loops should be avoided. The purpose of this section is to de-mystify the issue of grounding and show that there are bad *AND good* ground loops!

A good wiring practice for a voltage-input channel is shown in Figure 4.5(a). For this case, the sensor is not grounded and the cable shield is connected to the midpoint of the sensor as well as to the "ground" connection at the data acquisition system. This shield connection also meets the requirement of most instrumentation front-ends—*there must be a return path to the instrumentation common so that the input current (as low as it is) will not cause the input pair to "float" outside the common-mode range.* Failure to have this return path is a common cause of DAQ system problems. This will generally cause data to be collected with very poor linearity.

The situation is more complex if the sensor is grounded. The customary recommendation in this case is to connect the shield at the sensor and not at the DAQ system input *to avoid a ground loop*, as is shown in Figure 4.5(b). Another possible method to avoid a ground loop is to leave the shield unconnected at the sensor, as is shown in Figure 4.5(c). For most applications, the circuit shown in Figure 4.5(d) is recommended, with the shield connected at *both ends*, even though this creates a ground loop. This is the first example of a *good* ground loop. In nearly all cases, this double grounding has been shown to give far superior performance than any configuration with the shield "open" at either end.

The photographs in Figure 4.5 show the results of a test with the sensor ground and the instrumentation front-end ground derived from separate outside ac power drops to increase the noise voltage between these "grounds." A 10-ohm resistor was used to simulate the sensor. The measurements were taken from the output of an instrumentation amplifier with unity gain. For the ungrounded sensor, as shown in Figure 4.5(a), very low noise is present at the instrumentation amplifier output. Figure 4.5(b) shows the result when the shield is connected only at the sensor end. The noise level reached 500 millivolts. Similarly, Figure 4.5(c) shows a peak-to-peak noise level of 550 millivolts with the shield connected only at the amplifier end. Finally, when the shield was connected at *both ends*, creating a *good* ground loop, the peak-to-peak noise was reduced to 5 millivolts, as shown in Figure 4.5(d). Grounding at *both* ends reduced the noise input by a factor of about *100 to 1*. This configuration contains no filtering. If a single-pole 5 kHz low-pass filter is added, the noise is less than 1 millivolt when both ends are grounded.

How can it be that a ground loop substantially improves performance? A ground loop is generally *bad* if it involves a signal-carrying conductor. An example of this was shown in Figure 4.1(b), where the voltage produced by the ground current translated one-to-one into normal-mode voltage because the shield is a "return" for the signal. A ground loop is often *good* if it does not involve a signal-carrying conductor. The high level of noise was seen when the shield was connected at one end is primarily high-frequency "hash" that entered the system through reduced common-mode rejection and nonlinearities in the instrumentation amplifier at high frequencies. The cable capacitance and other factors greatly reduce the transmission of this noise when the shield is connected to the circuit elements at both ends. Indeed, with a connection to ground at both ends, current flows through the shield, particularly at power-line frequencies, and the signal conductors act as secondary windings of a transformer with the shield as the primary. The effect of this transformer action is greatly reduced because these voltages cancel out in the differential-input instrumentation amplifier. The double grounding cannot be applied if there is a substantial potential difference between the two circuit commons. For this case,

4.1. PROPER CONNECTIONS TO THE SENSORS

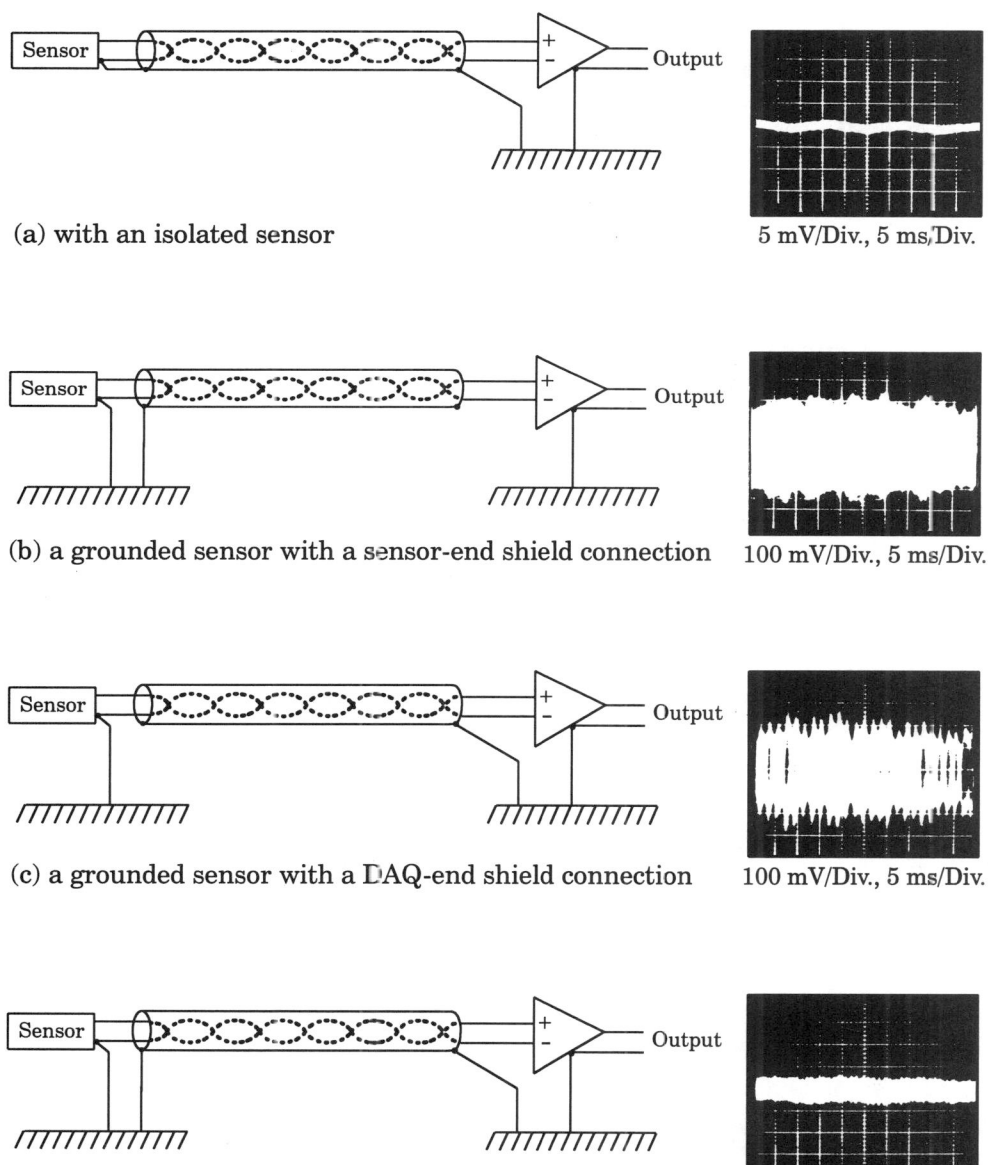

Figure 4.5: Noise levels with various shield grounding methods

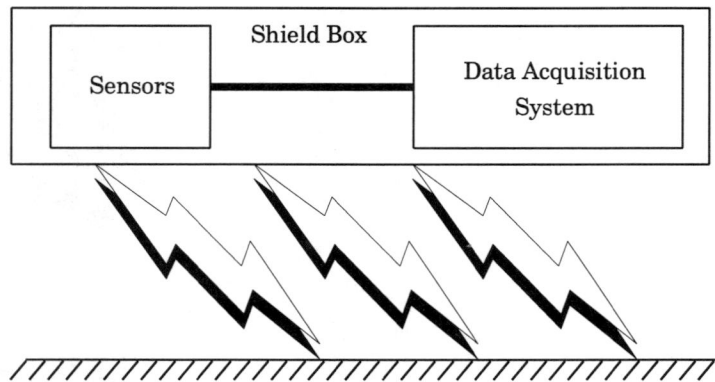

Figure 4.6: A data acquisition system at 1 million volts above ground

an isolated input channel, as shown in Figure 4.4, is appropriate. Generally speaking, if a "grounded" sensor and "grounded" instrumentation input produce a ground potential difference that prevents the grounding of the shield at both ends, input isolation must be provided.

Keeping the ground system as unipotential as possible is another very important aspect of reducing noise pickup in a data acquisition system. KineticSystems has supplied data acquisition chassis for research laboratory Van de Graff generators that are operating at 1 million volts above ground and are transmitting digital data to ground-connected hardware via fiber optics. Refer to Figure 4.6. How do these systems operate without excessive noise pickup? Just as people don't notice that the earth's surface is spinning at speeds up to 1,000 miles per hour because all of their surroundings are moving at the same speed, a data front-end and its sensors that are at the same potential and well shielded from a noisy environment can perform quite well. Note that the important factor is that the sensors, the wiring and the instrumentation front-end are at nearly the same potential.

If the data system uses more than one equipment rack, these racks should be bonded to each other directly by screws or by a large conductive strap. The rack system should be connected to a good ground—often electrical conduit. If there are wire ducts or conduit carrying the signal cables, the usual recommendation is that these be bonded to the ground reference for the sensors *and* the racks that contain the data front-end. Note that this creates another ground loop, usually the *good* kind. This approach is controversial when the sensors are in a rather hostile electrical environment. The concern often expressed is that ground bonding at both ends will cause the electrical noise at the sensors to be transmitted to the data system ground and create more interference. Generally, the noise

4.1. PROPER CONNECTIONS TO THE SENSORS

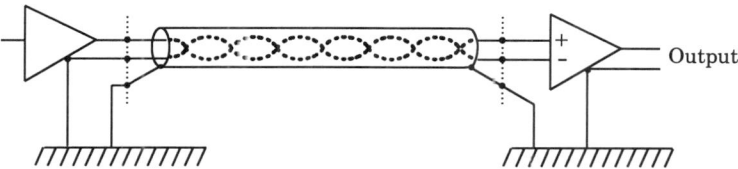

Figure 4.7: Driving a differential input channel from an unbalanced source

reduction resulting from the sensors and front-end being at nearly the same potential will far outweigh the introduction of noise by the ground loop. An additional benefit of ground bonding is that it will reduce the chance of damage to the electronics in the presence of lightning or other voltage transients. The lower the electrical impedance between the various parts of the circuit, the lower the potential difference in the event of a large voltage transient.

Even if the sensors are not connected to any part of an electrical circuit or ground, and the connection is as shown in Figure 4.2(a), the preferred technique is to bond the grounds at the sensors and DAQ front-end, using the input cable conduit, wire tray, or a # 8 AWG or heavier wire. This will minimize the effect of any electrostatic coupling of noise to the sensors. Again, this is to keep the entire data acquisition front-end, including the sensors, in as unipotential an environment as possible. A similar approach involves the use of double-shielded cables, where the inner shield is connected as in Figure 4.2(a) and the outer shield is connected to ground at the sensors and to the front-end equipment chassis ground. Again, the guiding principle is to keep *all* parts of the analog subsystem moving at the same potential, just like associated objects are spinning together on the surface of the earth.

Other approaches can be used if the primary common-mode (signal-to-ground) interference is primarily high frequency in nature. One configuration uses a trifilar transformer, which is a tightly coupled three-winding transformer, with one winding in series with each of the two signal conductors and the third winding in series with the guard connection. This is quite effective in enhancing common-mode rejection at high frequencies. Another approach is to use a capacitor to connect the circuit ground to the shield at the instrumentation front-end. This provides a high-frequency ground while reducing the current caused by power-line frequencies. The effectiveness of the capacitor depends upon the particular situation. Also, some front-ends provide a guard signal for connection to the shield. The guard voltage is derived from a special instrumentation amplifier output that monitors the common-mode voltage and produces a signal to cancel it.

Another question that is often asked involves the correct cabling and ground connection

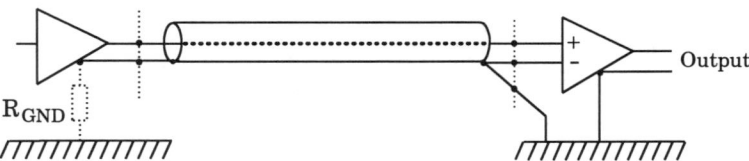

(a) The source is "floating" or has a high resistsnce path to ground

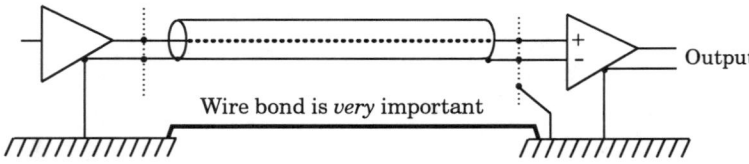

(b) The source is grounded

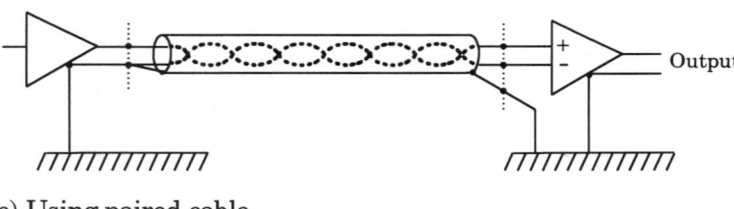

(c) Using paired cable

Figure 4.8: Connecting drivers with coaxial outputs to differential inputs

when the signal source is an amplifier output instead of a passive sensor. If properly wired, this output circuit can be single-ended and produce a good signal-to-noise ratio. If the amplifier output circuit drives a two-contact-plus-shield-type connector, the connections are shown in Figure 4.7. This provides a good balance because the output impedance of an instrumentation or operational amplifier is generally less than 1 ohm at low frequencies.

Some signal conditioning chassis contain BNC single-contact connectors at their output and are intended to be used with coaxial cable. This can present problems in correctly wiring a coaxial cable to a differential-input front-end. If the output shield connection on the signal conditioning unit is isolated from ground or has a resistance path to ground of 1000 ohms or greater, then the connections shown in Figure 4.8(a) should be used. If the shield conductor is grounded at the source, then the diagram shown in Figure 4.8(b) can be followed to prevent a ground loop in a signal-carrying conductor. This approach may not be satisfactory unless the data acquisition system contains a high-frequency filter. Also, it is *very* desirable that the ground frames of the chassis associated with the signal transmitter and receiver are mounted in the same rack or nearby racks and are electrically bonded together so that the ground noise between them is minimized. A good wiring alternative, particularly if the two units are not in the same rack, is to convert the cable to a shielded-twisted-pair type *as close to the source as possible* as shown in Figure 4.8(c).

4.2 Noise

The presence of noise in data can render that data meaningless. In some cases, data can be recovered through post processing. However, once noise has been added to the data, information is generally lost. Some of the sources of noise are:

- low frequency thermal drift,
- line frequency interference,
- induced RF signals, and
- harmonic distortion.

Every effort should be taken to eliminate the generation of noise, rather than to eliminate it later by filtering or other techniques. Most system noise problems can be solved by good installation practices.

Noise may be introduced external to the data aquisition system, or it may be induced by the system itself. Some sources of external noise are:

> *sensors* improper installation of the sensor
> *cabling* improper shielding and grounding (see Section 4.1.4)

One of the primary sources of noise within the data acquisition system is the instrumentation amplifier (IA) itself. This is true because the input to the IA is interfaced to the sensor and its wiring, and the lowest signal level in the system is usually at this amplifier's input terminals. Careful consideration should be given to matching the instrumentation amplifier to the system needs. For example, the error sources are worse for higher speed amplifiers. Therefore, the amplifier's bandwidth should not greatly exceed that required for the scan rate used. With wider amplifier bandwidth, dc performance is poorer and generated noise is greater.

One common problem is caused by incorrect allocation of gain between amplification stages. In general, whenever there are multiple gain stages in the signal path, the gain should be maximized in the first stage of an amplifier-filter combination. This has the effect of reducing the noise contribution of the later amplifier stages. When the signal path includes a low-pass filter, pre-filter gain should be maximized. The limiting condition is that the signal, with all of its frequency components, must not overload the pre-filter amplification stage(s).

4.3 Calibration

Historically, the process of calibrating a channel has been a time-consuming process. With a short circuit at the input connector, a potentiometer would be adjusted until the channel output read zero volts. This sets the channel's zero offset. Then the gain would be adjusted similarly, with a known voltage applied to the input. This process would need to be repeated later if a different gain range was required. The only other approach was to use components in the signal path which were extremely accurate and exhibited the lowest drift possible.

With the advent of modern data acquisition systems, the whole process can be automated to the extent that is feasible. Rather than adjusting the channel for a 'perfect' gain and offset, a better approach is to measure and record in memory the transfer function of the channel and then compensate, generally in software, for any deviations from the ideal when the data is being reconstructed. Thus, under computer control, the ADC count corresponding to zero volts and a voltage near full scale can be measured and the transfer function calculated. An on-board, precision, programmable voltage source is often used to facilitate this measurement.

One obvious advantage to programmable calibration, in addition to dramatically shortening setup time, is reduced cost, since only the calibration source needs to have a high degree of absolute accuracy. Components used in the signal path need to be appropriately linear and have sufficient stability between calibration runs.

4.4 Gain

In order to use the correct gain, it is necessary to determine the largest amplitude signal which will be seen at the channel input terminals. The appropriate gain is that which maps this voltage to nearly the full scale voltage of the ADC. For example, if the largest input voltage from the transducer is ± 10 millivolts, and the full scale input to the ADC is ± 10 volts, then a gain of 1000 would maximize the use of the ADC and provide the best resolution. It is generally necessary to use an amplifier gain that would result in a voltage below full scale so that any drift in the amplifier or ADC will not cause an overrange condition.

4.5 Auto-Ranging

Auto-ranging is a method for dynamically adjusting the gain of a channel to provide the optimum amplification for each measurement, rather than using a fixed gain based on the largest expected signal.

This technique is best suited to relatively low speed measurements, since it requires that two readings be taken for each data point. The first reading is used to determine the appropriate gain range. The gain is then switched to that range and the data is allowed to settle before the final reading is taken. This ensures that the ADC has the maximum resolution for any given reading.

4.6 Signal Conditioning

4.6.1 Bridge Conditioning

Excitation Supply

Most bridge transducers require dc excitation. Either constant-voltage or constant-current sources can be used for this requirement. The most common excitation source involves constant voltage and will be discussed here.

Critically important to the accuracy of bridge measurements is the stability of the excitation voltage. Where moderate-to-long cable runs are encountered, monitoring this voltage at the sensor allows the power supply to correct for voltage drops due to lead length and for resistance changes in the leads. The excitation supply should have sufficient closed loop bandwidth to correct for any noise induced in the leads. Therefore, this bandwidth should at least equal the measurement bandwidth.

In order to avoid possible damage to the transducer elements, the excitation supply voltage should decrease immediately if either an overcurrent or open sense-line condition is detected. The excitation monitor circuit also should be a source of a system alarm. In order to meet these requirements, one excitation power supply for a number of channels is generally inadequate. A separate excitation source per channel provides far superior performance. The ADC subsystem via software should be able to measure the excitation voltage directly at the transducer by monitoring the sense lines. This provides an accurate calibration of the bridge excitation.

Bridge Completion

Unless a full-bridge type of transducer is being used, bridge completion resistors will be required. These must be selected very carefully to ensure that they do not add errors to the measurement.

By its nature, a Wheatstone Bridge amplifies errors as well as signals. Therefore, any drift in the bridge completion resistors will be amplified and will degrade the measurement. Resistors with extremely low temperature coefficients, such as the Vishay metal-foil type, should be used. Signal conditioning modules that allow these resistors to be switched-in using low thermal EMF relays greatly reduce system setup time.

Shunt Calibration

Automatic insertion of shunt calibration resistors dramatically reduces the time required for the calibration process. This shunt calibration can be performed in such a way that it is not affected by lead length. This is an extremely important system characteristic.

4.6.2 Thermocouples

The most critical part of thermocouple measurement is the so-called "cold-junction" connection. If this is not handled properly, the errors created can reduce system accuracy dramatically. Since any junction between dissimilar metals will create a thermal EMF, the voltage generated by a thermocouple is the sum of all junctions around the loop. The drawing on the left in Figure 4.9 shows that there are typically three junctions for a thermocouple connection. Points A and C represent the connection of the thermocouple wires to the terminal block, with temperatures of T_A and T_C, respectively. If $T_A = T_C$, then the voltage read across the terminals becomes the contribution of the thermocouple at T_B minus the contribution of the thermocouple at T_A. However, if the two terminal temperatures are not equal, error terms remain.

An isothermal block is usually used beneath the terminal blocks to ensure a low temperature difference between these points. Additionally, air flow and heat sources should be eliminated from this area. This block also includes a temperature-measuring sensor, such as an RTD (a temperature-dependant resistor), that is read by the system for cold-junction compensation.

The accuracy with which the terminal temperature is measured obviously affects the overall accuracy. Since the transfer function of a thermocouple is nonlinear, a 1% error in measuring this temperature can result in an overall measurement error of several degrees.

Using an external "ice bath," as shown on the right in Figure 4.9, removes the need for measuring the cold junction temperature at A and C, but still requires an isothermal block at these junctions. The purpose of an "ice bath" is to keep this second thermocouple at a

4.7. READING THE ANALOG SPECIFICATIONS

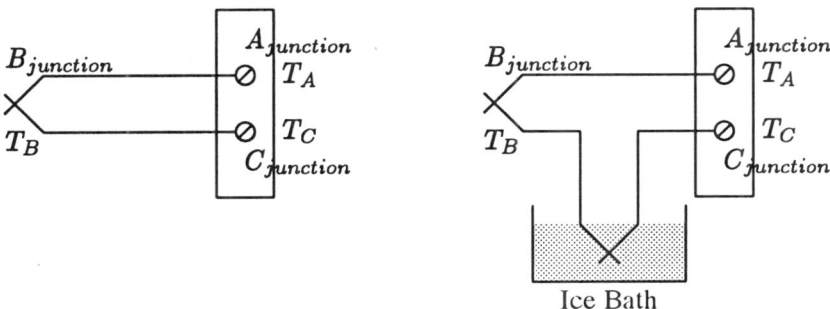

Figure 4.9: Typical thermocouple configuration

known temperature and input that value (about 0 degrees C for actual ice). The ice bath can be replaced by a temperature-controlled panel. In fact, the isothermal panel described earlier can use a temperature-controlled heater to keep the panel at a known temperature. This value is used by the system to determine cold-junction compensation.

4.7 Reading the Analog Specifications

Proper understanding of the data acquisition system's analog specifications is an important step in selecting and using the system. Certain elements of the specification are common to most applications. For example, consider a module that is needed to condition strain gage outputs. A sample data sheet is shown in Table 4.1. Each parameter will be discussed as well as its effect on overall measurement accuracy.

4.7.1 Inputs

For discussion, the assumption is that full scale ranges of ± 10mv to ± 30mv will be needed to handle the maximum excursions of the strain gage. At these low levels differential inputs are mandatory. The various sources of noise would mask the real signal if single-ended inputs were used.

The input impedance of the channel forms an attenuator when combined with the impedance of the source, creating a gain error. If this error is constant then it may be compensated during calibration. Unfortunately, the input impedance can be strongly affected by humidity, due to contaminants on the printed circuit card. If proper guarding and cleaning procedures are not followed, the input impedance can vary dramatically producing errors of tens of microvolts.

Table 4.1: Typical Specifications for Bridge Completion Signal Conditioning

Item	Specification
Inputs	
Number of Channels	Eight differential channels
Impedance	20 MΩ minimum; >100 MΩ typical
Overload Protection	To ± 20 volts, continuous
Gain Ranges	1, 2, 5, 10, 20, 50, 100, 200, 500, and 1000
Frequency Response	
Filter Type	6-pole Bessel or Chebychev
Cutoff Frequencies	20, 200, 1000, or 2000 Hz programmable
	The filter may be bypassed for an extended frequency response to a -3dB point of 20 kHz.
Excitation	Independent excitation for each channel. Each channel provides $\pm$excitation and sense leads. Excitation voltages of 0, 2.5, 5, 10, and 15 volts are available. Open sense lines or an over-current condition shut down the supply automatically and signal the mainframe of the error condition.
Bridge Completion	Eight channels of bridge completion are provided. 1/4-, 1/2-, and full-bridge configurations are supported. A matched pair of 120Ω or 350Ω resistors are provided for 1/2-bridge completion, while a third resistor may be used for 1/4-bridge completion.
Shunt Calibration	$\pm$ shunt calibration is performed on each channel. The customer-supplied resistors are installed in the termination assembly. Switching is performed under software control.
Gain/Offset Accuracy	After automatic calibration; see table below:
	Gain Accuracy
	1 $\pm$(1 mv + 0.02% of reading)
	2 $\pm$(1 mv + 0.02% of reading)
	10 $\pm$(0.1 mv + 0.03% of reading)
	500 $\pm$(0.02 mv + 0.1% of reading)
	1000 $\pm$(0.02 mv + 0.1% of reading)
Gain Stability	Better than 20 ppm per °C (typical)
Offset Voltage Stability	Less than 2 μvolts per °C RTI at gain of 1000
Noise	Less than 5 μvolts RTI at gain of 1000
Linearity	Better than 0.005%
CMRR	Typically better than 110dB, dc to 120 Hz at a gain of 1000. Trifilar-wound inputs provide excellent RF rejection to 100 MHz.
Bridge Balance	A 12-bit DAC provides the ability to remove bridge offsets of up to ± 70 millivolts.
Environmental	
Temperature Range	
Operational	0 to +50°C
Storage	-25°C to +75°C
Relative Humidity	0 to 85%, non-condensing

This impedance specification must be guaranteed over the temperature and humidity range to which the module will be exposed. The higher the input impedance, the less effect it will have on the desired data.

Overload protection of the input path should be adequate to handle any extraneous voltage to which the input may be subjected. An important note here is that proper ESD (electrostatic discharge) procedures, such as wrist ground straps, should be used whenever handling any printed circuit card in the system. Static discharge can destroy components on any PC card. Properly designed cards provide a reasonable degree of ESD protection at the channel input terminals.

4.7.2 Gain Ranges

It is necessary to have flexibility in selecting the appropriate gain for different tests. The goal is to be able to map the full scale span of the transducer to as high a percentage of the full scale range of the ADC as practical, providing maximum resolution of the signal. For example, if the full scale output of the transducer is ± 15mv and the ADC has a span of ± 10volts, a gain of 500 would use most of the ADC's range. Programmable gain selection allows the system to easily accommodate a variety of input conditions.

4.7.3 Frequency Response

Filter Type

The filter which best fits the application should be chosen. An important note is that, for time domain analysis, the smooth phase response of the Bessel filter usually will provide the best results. If overshoot and settling time are not important, then Chebychev or Butterworth filters are often used. These filters provide faster rolloff in the frequency domain for the same number of poles than a Bessel filter. This allows the sampling frequencies to be closer to the filter cutoff frequency. Increasing the number of poles in a filter increases the overall rolloff rate. However, this also increases the cost of the system.

Cutoff Frequencies

In general, the band of frequencies from zero to the cutoff frequency allow the information of interest to pass through the filter with low attenuation. Undesirable frequency components and noise will be attenuated above the cutoff frequency. One must remember that the higher the cutoff frequency, the higher the sample rate must be to prevent aliasing. This usually has significant system cost implications, particularly if a large number of channels are involved.

Collecting some data with the filter bypassed in order to estimate the spectral content of the signal is sometimes helpful. While this data may not be totally accurate because of aliasing, this test can be used to select the optimum cutoff frequency. If there is very little signal present at the higher frequencies, the filter does not need to provide as much attenuation at those frequencies. This test may allow sample rates to be reduced.

4.7.4 Bridge Completion

Flexible bridge completion with highly stable resistors is a desirable feature.

4.7.5 Shunt Calibration

The ability to automatically insert shunt calibration resistors across the bridge significantly shortens setup time and provides an end-to-end calibration of the channel. The minimum requirement for this function is the ability to insert a shunt calibration resistor across one arm of the bridge. Increased calibration accuracy can be obtained by switching the resistor in sequence across adjacent legs of the bridge, covering positive and negative swings of the transducer. The value of the calibration resistor is chosen to approximately simulate the maximum sensor excursion.

> **In large channel applications,** *such as automotive crash testing and structure load testing, a manual shunt calibration can often require many hours of pre-test setup time. However, KineticSystems engineers showed one company how to implement an auto insertion approach which reduced setup time from hours to minutes. The approach also facilitated the software design for archiving all important calibration data with the actual test data for later retrieval/verification/analysis.*

4.7.6 Accuracy

Overall system accuracy is the result of all components of the system, including the transducer. If the inherent accuracy of the transducer is 1% this will be the primary source of error if used with a data acquisition system that maintains, for example, a 0.01% accuracy. The accuracy of the signal conditioning circuitry generally matches that of the ADC.

4.7.7 Gain Stability and Offset Voltage Stability

Once a channel has been calibrated, stability is critical. Variation of the temperature by five degrees Celsius during the course of a test is not unusual. If the signal conditioning circuitry or the ADC have poor temperature stability, the accuracy of the data will be adversely affected.

Regardless of how high the circuit stability is specified by the system manufacturer, these values usually assume that there is sufficient warm-up time before calibration or data measurement to allow the the components to stabilize. Typically, about 15 minutes of warm-up is adequate.

4.7.8 Noise

This specification covers noise that is generated within the signal conditioning hardware. Generally, this parameter is specified at high *gains* where noise is most likely to be significant. In general, a wideband circuit will have higher noise than one with lower bandwidth. The design of the signal path involves a tradeoff between dc parameters such as drift, noise and linearity, and ac parameters such as slew rate bandwidth and harmonic distortion. Therefore, it is important not to overspecify any of these parameters.

4.7.9 Linearity

Often referred to as *integral linearity*, this is an error term in the transfer function of a channel which cannot be corrected by calibration. This value refers to the maximum deviation of any point from a best-fit straight line. To achieve good linearity, the circuit designer chooses operational amplifiers with high open loop gains. Unfortunately this characteristic of the amplifier parallels bandwidth, requiring that a tradeoff again be made. Linearity is also affected by the excess loop gain of an amplifier stage. Therefore, not all of the gain for a channel is placed in a single gain stage. A good guideline is to limit the gain of any one stage to 100 or less.

4.7.10 Common Mode Rejection Ratio

Common mode rejection ratio (CMRR) was discussed earlier in this chapter as it relates to sensor connections. For a low-level front end such as a bridge signal conditioner, the common mode rejection ratio should reach 100 dB at high gain. This means that the ratio of the normal mode gain to the common mode gain should be at least 100,000 to 1. This rejection should extend from dc to 120 Hz, since much of the common mode noise will be generated by line frequency and its first harmonic.

Studies have shown that common mode noise in a typical laboratory or industrial environment contains significant energy even at RF frequencies. These high frequency signals, well beyond the bandwidth of the instrumentation amplifier, may be rectified in the input stage of the amplifier and create a dc voltage shift which varies as the common mode noise changes. This can create significant measurement errors in a system. It is therefore desirable to attenuate these RF common mode signals before they reach the input amplifier. This can be accomplished by using a trifilar-wound input transformer, which provides an extremely high impedance for high-frequency common-mode signals, but does not affect the normal-mode signal.

> **A leading aviation company** *was looking for a high performance data acquisition system and had established its own specifications. Unable to find a suitable signal conditioner in the marketplace to meet the requirements, they contacted KineticSystems Corporation. One of the critical design parameters, and the customer's primary concern, was the handling of RF common mode signals. KineticSystems engineers, working with the customer, designed an approach using a trifilar wound input transformer. This helped the designers meet their needs and eliminate error source for their environment.*

4.7.11 Bridge Balance

This is a method that allows a bridge to be automatically balanced, usually by injecting current into it via an on-board Digital-to-Analog Converter (DAC). The bridge balancing may be accomplished automatically by the signal conditioning hardware or may involve supporting system software. A balanced bridge can be important because it generally allows a higher channel gain to be used.

Chapter 5

Filtering

Filtering is used in data acquisition for several reasons:

- To limit the input signal bandwidth to prevent signal aliasing

- To eliminate noise, particularly high frequency

- To eliminate stray pickup, e.g., 50 or 60 Hz

A common reason for filtering is to eliminate signals with frequency components above the Nyquist frequency that would be aliased to lower frequencies. In the case of antialiasing, the filter is a lowpass or bandpass filter, and the filtering *must* occur before the signal is sampled or multiplexed. Thus, filtering is *always* performed on a per-channel basis. Also, if a low-level signal is involved, amplifying the signal prior to filtering is generally necessary to minimize the effects of noise. These two considerations both tend to increase the cost of solving the aliasing problem, since taking advantage of common circuitry after multiplexing several different signals is not possible.

Filtering to limit high frequency noise is another common requirement, and like antialias filtering, it must be implemented on a per-channel basis prior to any multiplexing or sampling. Noise filtering is usually less stringent than antialiasing since the noise floor is typically low, and the filter simply limits the high frequencies.

Filters are not without their own issues. They introduce their own kinds of distortion, and matching their responses across a number of different channels can be expensive, especially when high roll-off filters are needed to minimize the sampling rate. Section 5.2 discusses the various *classic* types and applications of filters.

5.1 Antialias Filter Considerations

Designing a system with antialiasing filters involves a number of very critical considerations. These include:

- The system cost (exclusive of filters) for a given sampling rate.

- The filter cost, particularly as sharper filters are used to allow a lower sampling rate.

- The channel-to-channel phase and gain matching required.

- The phase and gain response of the "type" filter needed.

- The frequency characteristics of the signal being acquired.

It is important to realize that *all the above considerations have a direct impact on the cost of the overall solution* and *that they are tightly coupled to each other*.

5.1.1 System Costs

System cost, exclusive of the antialiasing filters, is typically driven by sampling rate or throughput and accuracy requirements. With today's technology, system cost scales fairly well with sustained average system throughput to approximately 500,000 samples/second, and then rises fairly sharply due to limitations on throughput of buses and storage devices. When considering system cost, higher sampling rates mean larger memories, larger disks, higher bandwidth buses, increased processor power, and higher analysis costs (more data to process). A frequently overlooked consideration is that more data may involve more staff time to process, analyze, and interpret the data.

Overall system accuracy is the other cost driver. Assuming a "16-bit" ADC gives you 16 bits of accuracy is not sufficient. Refer to Section 6.2 for more details. Overall system accuracy starts with the sensor, and involves *every component*, up to and including the ADC. Also, system accuracy is typically different at low frequencies than at frequencies near the top of the pass band. If one wants *true 16-bit accuracy*, one must be prepared for a sizable investment and look very carefully at the specifications.

KineticSystems Corporation *is the only resource you need for building high performance data acquisition and control system solutions.*

Software or hardware, KineticSystems has a full array of products to meet your needs, whether you use a PC or a workstation.

900 N. State Street
Lockport, IL 60441-2292

Building System Solutions...
Together.

Our products include:

- Complete data acquisition systems.
- Over 300 VXI and CAMAC products, including sophisticated signal conditioning.
- H•TMS™: a totally integrated, UNIX-based data acquisition system.
- Reality™: a state-of-the-art distributed realtime data acquisition and control software package for UNIX-based workstations.
- KSCrti (formerly DEC Realtime Integrator™): an icon-based software for VAX/VMS and Alpha VMS/OSF.
- Software drivers for LabVIEW® and other popular data acquisition software packages.

Contact us today to discuss your specific test and measurement application.

Phone: 1-800-DATA-NOW
        (1-800-328-2669)
Fax: (815) 838-4424
E-Mail: SALES@KSCORP.COM

Yes, please send FREE information!

Send me free information on
(check all that apply):
- ☐ VXI products
- ☐ CAMAC products
- ☐ H•TMS™ systems
- ☐ Reality™ software
- ☐ KSC Realtime Integrator Software
- ☐ LabVIEW® software drivers
- ☐ Other _____

My interest is:
- ☐ immediate   ☐ 3-6 months
- ☐ 6-12 months   ☐ reference only
- ☐ **I am interested in a FREE demonstration.**

Comments:

Fill out and return this card today.

Name
_____
Title/Company
_____
Street Address
_____
City/State/Zip
_____
Phone
_____

You should also contact:
Name
_____
Title/Company
_____
Street Address
_____
City/State/Zip
_____
Phone
_____

Code: 3054

Building System Solutions...
Together.

900 N. State Street
Lockport, IL 60441-2292

Contact us today to discuss your specific test and measurement application.

Phone: 1-800-DATA-NOW
         (1-800-328-2669)
Fax: (815) 838-4424
E-Mail: SALES@KSCORP.COM

No Postage Necessary If Mailed In The United States

BUSINESS REPLY MAIL
FIRST CLASS MAIL PERMIT NO. 109 LOCKPORT, IL
POSTAGE WILL BE PAID BY ADDRESSEE

KINETICSYSTEMS CORPORATION
900 N STATE ST
LOCKPORT IL 60441-9986

5.1.2 Filter Costs

Generally filters can be categorized by:

- Filter Type (Bessel, Butterworth, Chebyshev, Elliptical,...).
- Number of Poles (typically 2, 4, 6, or 8).
- Channel-to-channel gain match.
- Channel-to-channel phase match.
- Passband ripple.
- Stopband floor.

In general, filter cost is driven primarily by the number of poles, which govern the sharpness of a given type filter, and by the degree of gain and phase match near the cutoff frequency. The number of poles drives the amplifier stages needed to implement the filter and the gain-bandwidth of the operational amplifiers as well as the required precision resistors and capacitors. The gain and phase match, as well as the number of poles, drive the precision of the components needed to implement the filter.

5.2 Types of Filters

Lowpass filters provide a means for limiting the bandwidth of signals to be sampled and digitized. Depending on the nature of the application and type of analysis one plans to perform on the data, different filters offer various advantages. In general, one needs to select a filter that preserves the characteristics of the information that is of interest in the application.

This section reviews characteristics of the classical filter designs.

5.2.1 Bessel Filters

The Bessel filter provides a linear phase response with a ripple-free passband and monotonic rolloff. It delays signals at passband frequencies by a constant amount of time and has an overshoot-free step response. The Bessel filter produces a delayed (but accurate) replica of the input signal. These characteristics make the Bessel filter ideal for time domain applications.

The primary limitation of the Bessel filter is its relatively slow roll-off compared to other filters with the same number of poles.

5.2.2 Butterworth Filters

The Butterworth filter has a maximally flat frequency response in the passband with a monotonic rolloff that is much sharper than the Bessel filter. Its phase response varies non-linearly with frequency; the delay is no longer constant and the step response exhibits a moderate amount of overshoot (ringing). These characteristics present no problems for amplitude-based applications. The Butterworth filter is a good general purpose filter.

5.2.3 Chebyshev Filters

The Chebyshev filter provides an even faster rolloff than Butterworth, but exhibits an equal-amplitude ripple across the passband as well as a step response with more overshoot than the Butterworth filter. The phase response is nonlinear and the passband delay is not constant which results in serious ringing in the time domain. This frequency dependence may not present a problem when the primary focus is the attenuation characteristics needed to minimize oversampling and reducing sampling rates.

5.2.4 Elliptic or Cauer Filters

The elliptic or Cauer active lowpass filter has a wide and nearly flat passband response with an extremely sharp roll-off characteristic. It has equal-amplitude ripple in the passband (typically 0.1 dB) and a well-defined attenuation floor in the stopband. This filter also has a nonlinear phase response as well as an overshoot step response. The elliptic filter is ideally suited to amplitude-based antialiasing applications. When selecting an elliptic filter, the attenuation floor must be low enough to insure that aliased stopband frequencies are sufficiently attenuated to not introduce significant errors into the measurement. This is generally not an issue with other filters since they have monotonic rolloffs.

5.3 Filter Implementations

With today's technology, most high-performance filters are implemented as continuous active filters—or in the case of one- and two-pole filters, passive RC filters may be used. Switched capacitor filter technology is improving and is used in some less demanding applications.

5.3.1 Continuous Filters

Continuous active filters introduce minimum noise into the signal; however, it is not easy to provide switchable or programmable bandedges. Components need to be very accurate, especially in the high Q stages used for Chebyshev and Elliptic filter designs.

An efficient way to provide flexibility in filtering is to have a relatively small number of programmable analog bandedges (e.g. decade boundaries) to manage noise and antialiasing considerations, followed by digital filtering to provide exact cutoff characteristics where needed. Refer to Section 5.5.

> **In automotive safety tests,** there is a need to provide different classes of filtering to meet SAE J211 requirements. To accomplish this, KineticSystems Corporation combines analog anti-alias filters with digital filtering techniques. This provides a flexible cost-effective approach to meet specific filter characteristics.

5.3.2 Switched Capacitor Filters

The cutoff frequency of switched capacitor filters is determined by controlling the input clock frequency, allowing easy control of the cutoff or corner frequency. Unfortunately, the output of the switched capacitor filter must be filtered again with a continuous filter to reduce clock noise. If this post filter has a fixed bandedge, then obviously it can handle only a limited range of corner frequencies from the switched capacitor filter. The limiting factor in switched capacitor filters for high-performance applications is the noise floor that can be achieved in an environment with a very high frequency clock.

5.4 Choosing Filters and Sampling Rates

Clearly, the first decision is selecting the passband frequency range. This is determined by either the bandwidth of the transducer or the bandwidth of the information of interest. Related to selecting the filter is the frequency range of the signals that the filter will pass relatively unattenuated. For a lowpass filter this is characterized by the cutoff frequency f_c where the filter attenuates signals by 3dB. The choice of filter type may play a role as far as the amount of passband ripple (the variation in filter attenuation in the passband) that can be tolerated. Passband ripple affects system accuracy, particularly near the cut-off frequency.

The next decision involves choosing a filter having characteristics consistent with the kind of analysis one intends to perform (refer to Section 5.2). Generally, one would choose filters with the steepest rolloff for a given number of poles without compromising the data.

Finally, one must choose a sampling rate and a filter with an appropriate number of poles that will insure that any aliased signal will be sufficiently attenuated as to have an acceptably small effect on the information content of the data. This is far easier said than done, and the decision can have dramatic effects on the system cost.

Understanding of the frequency spectrum of the input signals is an important consideration. An expensive filter is not needed to deal with a signal that has negligible frequency components above the frequency range of interest. In this case, a relatively simple lowpass filter with a cutoff frequency f_c just above the frequency range of interest will suffice to limit noise and aliasing. A sampling rate consistent with the highest frequency of interest will suffice—typically 3 to 5 times f_c.

In the more general case, the object is to choose a filter with a sharp enough roll-off to insure that all frequency components of the signal above the Nyquist frequency (one-half the sampling frequency) are attenuated sufficiently so that their contribution to the overall system error is within acceptable limits. Further, the filter should not introduce artifacts (such as those caused by passband ripple) into the signals in the passband that result in an overly large system error.

5.4.1 ADC Quantization Noise

Another consideration is *how small must a signal be before it does not significantly degrade the overall accuracy of the measurement, or, more importantly, does not mask the information of interest?* Certainly the ADC resolution imposes one limitation. For example an N-bit ADC ($N - 1$ bits plus sign) has 2^{N-1} possible "discrete values." Assuming a perfect ADC, the maximum uncertainty of any measured value is $\pm 1/2$ least significant bit (lsb) or 1 part in 2^N. When such an ADC is used to digitize a continuous signal, the difference between the quantized—or digitized—value and the signal can be viewed as an error or noise (quantization noise). A more in-depth analysis is provided in the book by Oppenheim[6] which gives a signal to noise ratio (SNR) in decibels (dB) for an ideal ADC of

$$\text{SNR} \approx 6B - 1.25 \text{dB},$$

where B is the number of bits of resolution of the ADC. Attenuating the signal frequency components above the Nyquist frequency significantly below this value provides minimum improvement in overall system accuracy. Table 5.1 provides theoretical SNR values for some *typical* ADC resolutions.

5.4.2 Signal Characteristics

In choosing a filter/sampling rate combination, considering the frequency characteristics of the signal is important, specifically the frequency components above the filter cutoff frequency f_c, and particularly those above the Nyquist frequency, f_n. The *standard operating assumption* in the industry for choosing antialias filters is to *assume the signal*

5.4. CHOOSING FILTERS AND SAMPLING RATES

Table 5.1: Theoretical Signal to Noise Ratios for Typical ADCs

ADC Resolution	SNR	Limiting Resolution
12-bits	-70.75 dB	0.0290%
14-bits	-82.75 dB	0.0073%
16-bits	-94.75 dB	0.0018%

components above f_n are of equal strength as in the passband. This assumption is probably better for system suppliers than users, as it generally leads to an overly conservative system configuration.

More frequently than not, the major portion of the signal strength is in the passband. If not, then it is possible that the interesting information may lie at higher frequencies and the f_c should have been set higher. For example, if all of the high-frequency signal strengths falls below 10% of the passband frequency strength, then there is 20dB less attenuation needed from the filter to attenuate frequencies above f_n. This translates into a much less expensive filter—or a lower sampling rate.

5.4.3 Selecting the Filter and the Sampling Rate

The various issues for establishing a sampling rate and filter have been discussed. The most practical procedure for reaching a final selection is unfortunately iterative.

1. The desired accuracy of the measurement must be established, but *not lower than the limiting value for the ADC given in table 5.1*. Making this error percentage artificially small will increase the system cost.

2. The type of filter (Bessel, Butterworth, ...) that best meets your needs must be established. One may need to compromise here to get the roll-off one needs to keep sampling rates down. Also, one will need to consider passband ripple and stopband attenuation as well as phase response and step or impulse response of the filter type selected. Refer to Section 5.2.

3. A cutoff frequency f_c for the filter must be established. The larger one makes this, the higher the sampling rate the system will need.

4. The frequency characteristics of your signal must be determined. How much is the signal strength down at two or three times f_c and above? Some reasonable assumptions should be made here. The better the ratio of passband to stopband signal strengths, the less the filter will need to do.

5. The amount of filter attenuation one needs to attenuate the stopband signals (Item 4) to a value that is below the measurement accuracy (Item 1) must be computed. For example, if the ratio of passband to stopband signal strength is 10:1 or -20dB (Item 4), and the desired accuracy is 0.1% or -60dB, then the filter needs to give 40dB attenuation for frequencies above the Nyquist frequency f_n.

6. The attenuation curves for the filter that one has chosen must be examined to determine the frequency above which the filter attenuation meets or exceeds the attenuation established in item 5. For a 6-pole Butterworth filter, an attenuation of -47dB is achieved at $2.5f_c = f_n$. Thus, in this example, a sampling frequency of $2f_n = 5f_c$ is required, and a 12-bit ADC will provide the required accuracy (Refer to Table 5.1).

Note that we could repeat items 5–6 for filters of the same type but having a different number of poles to optimize system costs. For example, an 8-pole Butterworth filter would reduce the sampling rate to $4f_c$. One could change filter types and repeat items 2–6. For example, an 8-pole Elliptic filter would reduce the sampling rate to $3f_c$.

In some applications the data analysis is performed in the frequency domain, i.e., the time domain data is first processed using a Fast Fourier Transform (FFT) or equivalent. In these cases, compromising further with the selection of the filter and sampling rate is possible. Rather than requiring the filter to attenuate signals at the Nyquist frequency f_n and above to the limiting accuracy, it is sufficient to set this requirement at the *highest frequency of interest* –typically near the filter cutoff f_c. This is illustrated in Figure 5.1 with an 8-Pole Butterworth Filter. Using this technique and the example in Figure 5.1, a sampling frequency of $3.8f_c$ is needed, as opposed to twice the frequency where the attenuation drops below the desired *accuracy floor*, or $5.6f_c$.

5.5 Digital Filtering

With the advent of low-cost Digital Signal Processors (DSPs) and digital filtering Integrated Circuits (ICs), it is now economical to consider the use of these components in realtime filtering applications. The issue that *must be kept firmly in mind* is that...

Solving the aliasing problem must be done **before** *one samples!*

Initially, it may seem that digital filtering doesn't help significantly, but in some applications with critical filtering needs—particularly where there is a need for tight gain

5.5. DIGITAL FILTERING

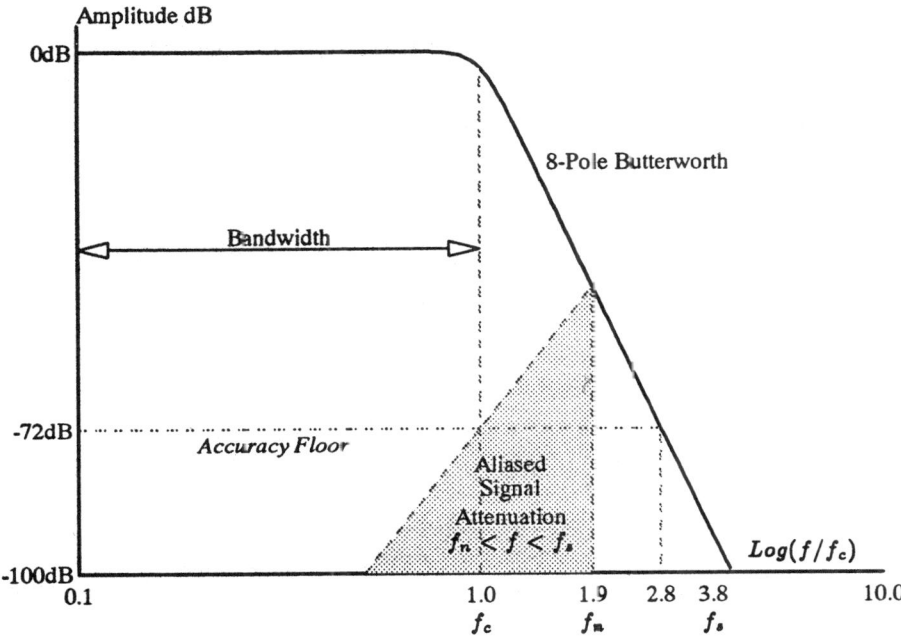

Figure 5.1: An 8-Pole Butterworth filter illustrating sampling rate set so aliased signals above accuracy threshold fall back to f_c for frequency domain analysis applications.

and phase matching requirements—digital filtering can be a cost-effective solution. The technique is to use lower-cost, slower-rolloff filters with relaxed gain and phase matching, and use a high speed ADC to oversample. Since the signals of interest are well below the filter cutoff frequency, there is considerably less sensitivity to the gain and phase matching. Also, a tradeoff exists in choosing the sampling rate, since allowing signals to alias into the front-end filter's passband is acceptable if the front-end analog filter removes any aliased signal between zero and the bandpass of the digital filter.

Finally, the digitized signal is passed to a DSP to perform the digital filtering and decimation (downsampling with antialiasing), resulting in a reduced output data rate. Since the filtering is now digital, channel-to-channel gain and phase matching are nearly identical to within the precision of the DSP, and implementing a wide variety of filters with custom characteristics is feasible. Further, it is possible to implement digital filters that are beyond today's analog technology.

By decimating the resulting filtered signal at the DSP, it is possible to reduce the effective sampling rate down to a value consistent with the passband characteristics of the digital filter and the sampling theorem. This reduces the bandwidth and storage requirements on the rest of the system.

One of the techniques that critically depend on this approach is the Sigma-Delta ADC. With this device, a very fast one-bit ADC is used to highly oversample the signal, and digital filtering techniques are used to "average" the sampled data to gain precision at a lower effective data rate. Refer to Section 6.1.5.

Chapter 6

Analog to Digital Converters

The Analog-to-Digital Converters (ADC), multiplexer (MUX) and sample-and-hold (S/H) amplifier form major elements of most data acquisition systems. It is important to understand the overall architecture of the ADCs and related components in order to choose an optimum system configuration.

6.1 Types of ADCs

The best type of analog-to-digital converter for a given application is determined by the requirements of that application. For example, if a slowly-varying thermocouple input is being measured, then the appropriate ADC might be of the integrating type.

6.1.1 Integrating ADC

This type of converter typically integrates over an integral number of cycles of powerline frequency and provides excellent attenuation of noise at that frequency and its harmonics. These converters may be of the dual-slope or multi-slope type. They provide a high degree of rejection at line frequencies and noise rejection, but have rather poor time resolution and speed.

6.1.2 Successive Approximation ADC

The most common type of ADC in use today is the successive approximation type. These converters combine high speed with high accuracy, have resolutions from 8 to 18 bits and conversion times from hundreds of microseconds down to submicroseconds. The higher speed ADCs typically have lower resolution. These ADCs are often used with a sample-and-hold amplifier. This element is necessary because the data must be held rather constant during conversion or significant measurement error may occur. The S/H amplifier must be carefully matched to ADC, with important parameters such as droop, aperture uncertainty, linearity and gain accuracy.

6.1.3 Flash Converters

Flash converters are usually very high speed with lower resolution. Such converters are generally used to acquire data with scanning rates from 5 Megahertz to 500 Megahertz. Since the flash converter uses one comparator per bit combination, these converters are often limited to resolutions of 8 bits (256-bit combinations).

6.1.4 Hybrid Converters

Many converters combine several techniques to optimize the resolution, speed and performance. Techniques include multiple-pass converters, error correction, etc.

6.1.5 Sigma-Delta Converters

Sigma-delta (sometimes referred to as delta-sigma) modulation was introduced in 1962, but it has only been recently that developments in VLSI technology have provided a practical means to implement the technology. A sigma-delta converter quantizes an analog signal with very low resolution (1 bit) at a very high sampling rate, typically in the 1 to 12 MHz range. The resultant signal is then passed to a digital lowpass decimation filter. The effect of this filter is to "average" the 1-bit samples, yielding a higher resolution (usually in the 16- to 18-bit range, but as high as 24 bits). This process results in an associated reduction in signal bandwidth and a much lower effective sampling rate. The high sampling rate is described as oversampling. Typically, a sigma-delta converter uses 64 times oversampling.

Sigma-delta technology offers several advantages over a conventional ADC for some applications. Due to the very high oversampling rate and digital filtering, antialias filtering is greatly simplified. A simple one-pole filter at the input usually suffices to attenuate the frequencies in the passband of the digital filter to well below detectable levels. Thus, issues of nonlinear phase can be avoided. Also, Sigma-delta converters exhibit better differential nonlinearity, greater dynamic range and improved signal-to-noise ratios over equivalent ADC subsystems.

6.2 Resolution and Accuracy

One of the more confusing tasks when evaluating data sheets is to separate the accuracy of an ADC from its resolution. Just because an ADC exhibits 16 bits of resolution does not guarantee 16 bit accuracy. In fact, an ADC may not even maintain monotonicity (each bit combination measuring a higher value than the next lower combination) over its full operating range, especially over the temperature range. In some applications, resolution is more important than accuracy; it may be more important to be able to resolve small changes in a variable than to know the absolute value of the new signal level. Many 16-bit resolution ADCs exhibit 14 bits of accuracy, and this is sufficient for a large number of applications.

In some applications absolute accuracy is paramount; therefore overall accuracy becomes more important than resolution. In these cases, it is important to look at the integral nonlinearity specification. This represents an error which cannot, in general, be eliminated with calibration.

Understanding the overall system requirements will aid in the selection of the appropriate ADC without overspecifying its characteristics. Higher cost is generally associated with ADCs that have more stringent characteristics. If dynamic accuracy is important, then it may be better to choose a faster, lower-resolution ADC than one with high resolution but longer conversion time.

6.3 Multiplexing Analog Signals

The analog to digital converter is often a relatively expensive component. In multichannel applications it is frequently more economical to multiplex many analog signals into a single ADC. At slower scan rates this can be very economical because the ADC, sample-and-hold amplifier and front-end gain stages can be shared by a large number of channels. Modern ADC modules of this type typically include a multiplexer, a programmable-gain amplifier and an ADC. The gain for each analog channel is stored in on-board RAM memory, and the gain is set individually for each input channel to the MUX. At higher scan rates, the settling time following a gain change and slew-rate limits of the amplifier section reduce the effectiveness of this architecture.

At intermediate scan rates, particularly in high gain situations, a separate amplifier must be provided for each channel. Also, in applications which require low-pass filters, these elements must be on a per channel basis and cannot be shared. At high sampling rates, settling times preclude the use of multiplexers. In such cases, it is more appropriate to use an ADC per channel. As was discussed earlier, sigma-delta converters may represent a better choice than multiplexing for some applications, even at lower scan rates.

6.4 Sequential Scan vs. Simultaneous Sample-and-Hold

In many applications, it is sufficient to sequentially scan and convert each channel with a multiplexed ADC. Using this approach, successive channels are sampled at consecutive time increments during each scan. The ADC may be self-scanning, or the scan may be initiated based on a system trigger. Self-scanning ADCs are generally appropriate for low sampling rate applications where "the most recent value of the signal" is acceptable—the exact time of the sample is unknown to within the scan interval, typically several milliseconds.

In more demanding data acquisition applications, it is important to sample data at evenly spaced intervals. Also, some time-skew between successive channels may be acceptable, as long as the interval between samples on any given channel is stable. In these applications, the ADC scan is triggered by a hardware clock. Typical aperture uncertainty (variation in sampling interval) for a given channel is approximately half the ADC clocking rate.

6.4.1 Simultaneous Sampling

In some very demanding applications, it is essential that the signals from *all* channels be sampled simultaneously and that the time-jitter in sampling of each channel be tightly controlled. In these cases, it is necessary to use a sample-and-hold (S/H) amplifier associated with each channel with all the S/H amplifiers clocked simultaneously. This approach typically reduces the aperture uncertainty or jitter into the sub-nanosecond range.

Chapter 7

Digital Processing and Buffering

One of the critical considerations in configuring a data acquisition system, involves moving the data from the ADC(s) or digital input device(s) into the computer for processing and generally saving the raw or processed data to some storage device. For control applications, the sequence is getting the data in, running the control algorithms, and outputting the control values. The initial and most obvious question in configuring systems is whether the data paths between the front-end to the processor, and the processor to the storage medium, will sustain the required throughput. What is often not adequately addressed are the sources and effects of latency and how they affect overall throughput. For example, a bus with 100 Mbyte/second throughput is useless for an application that needs to transfer 4bytes of data every $10\mu s$ (0.4 Mbyte/second throughput) if it shares the bus with a device that can acquire and hold the bus for $15\mu s$—or, if a processor is involved, the interrupt latency of the processor may exceed the time between samples.

There are many techniques for addressing these issues. Included are buffering, the use of dedicated processors, dedicated direct memory access (DMA) controllers, and configuring system I/O buses to conform to the application requirements. This chapter will explore some of the issues and techniques for addressing these problems.

7.1 Throughput

Throughput is a measure of the rate that data that can be moved from one point to another in a system and is typically measured in bytes per second. Most resources, such as disks, processors, and buses in a data acquisition system, are *shared* at some level among various tasks, processes, and/or devices. When throughput is specified, it is usually given for a dedicated service. Thus, when a bus has a specified or *theoretical* throughput of 40 Mbytes/second, the specification is for a dedicated transfer between two devices on the bus with the bus limiting the transfer rate. It assumes that no other devices are contending for the bus and that devices respond to bus transaction requests in minimum allowed times. Generally, this throughput is fairly easily measured or calculated.

> **NASA turned to KineticSystems Corporation** *when they wanted to use a variety of simulators in multisite simulation jobs. Because the realtime simulation system at the NASA Langley Research Center consists of 28 flight simulators located in several buildings, and the system needed to use powerful central computers, a low latency distributed I/O system with 3 megabytes per second throughput was required. KineticSystems Corporation supplied a CAMAC-based system with 55 I/O chassis distributed throughout the simulator site. With this system, a programmable highway switch allocates sites to computer channels in any combination and provides for up to 12 multisite simulation jobs at any one time. It includes thousands of channels of 16-bit analog I/O, and the I/O chassis are interconnected by fiber optic links operating at 50 megabits per second. This system exceeded NASA's requirements and received* Research and Development Magazine's *IR-100 Award as well as NASA's Space Award.*

7.2 Latency

Latency is a measure of the time it takes for a device or process to *initiate* a transfer of data from one point to another in the system. There are two components to latency, the *inherent* time required to initiate the action in an otherwise idle situation, and the time required to acquire and access shared resources. The first is determined by the design of the device or process being considered, while the latter is typically dependent on the detailed configuration and application at hand. In many cases latencies resulting from shared resources dominate.

Latency is frequently *much harder* to quantify than throughput, since, in many cases, obtaining a precise signal indicating when a request is initiated and when the transfer starts is not easily obtained. To complicate the issue, latency is very dependent on what other processes and/or devices are performing concurrently. Also, since most processes run asynchronous to each other, the effects of latency on a given process are usually statistical. Thus, to fully understand latency issues, making thousands of individual latency measurements is necessary and then histograming the results. In terms of quantifying the results, such statistical measures as *average latency* or in some cases *worst case latency* under a given (known) loading condition is sensible.

In any case, latency has two major effects: It limits the *realizable* throughput and the guaranteed minimum interval that a device can be serviced. In data acquisition systems, worst case latency determines the depth of buffers needed to prevent loss of data, and average latency determines the *realizable* throughput.

7.3. OPERATING SYSTEM LATENCY

In control systems worst case latency, along with throughput and control loop execution time, determines the minimum *guaranteed* time to close the loop. In many control applications it is frequently acceptable if the loop is closed "most" of the time within some specified interval. However, an occasional longer interval is acceptable if the loop is closed within some specified upper bound.

7.3 Operating System Latency

The *worst* latency offenders are, in fact, operating systems. The Operating System (OS), after all, is the arbiter of one of the most frequently shared resource elements, the CPU and its related resources such as memory, disks, network interfaces, and other I/O devices. The OS arbitrates between various applications and I/O processes within the computer system, and manages priorities between many concurrent processes as well as its resources. The data acquisition or control processes are just part of the larger picture. Even in a relatively dedicated application there are typically many related processes including I/O, data processing, data management, networking, and operator interface, each with its demands on a different set of resources under the OS.

The issue of interrupt latency is a relatively simple example. For typical realtime OS and processor, interrupt latency is usually under $10\mu s$ (average). Now consider a typical data acquisition application that is taking data, storing it to disk, and is likely connected to a network. Usually in such a scenario the worst case latency is several hundred microseconds to one millisecond, depending on the OS and the processor. This may seem excessive, but important factors are that the data acquisition interrupt may get delayed by the interrupt processing time of the disk, network, or system clock service routine; that the interrupt might be a result of some exception that requires extra processing at elevated priority; and that the system clock will usually have priority over any other interrupt and will almost always be serviced before your data acquisition interrupt.

A careful review of exceptional claims in this area is important. The computer sales business is *very* competitive and obtaining meaningful numbers from computer vendors is not easy. One can frequently get the "interrupt latency" but it's most likely the average for an unloaded system (best-case latency). Those who claim absolutely *super* numbers are using dedicated processors to achieve them (all the rest of the OS runs on another processor).

At the *process level* where one is writing relatively normal application code, latencies can be significantly higher. Figure 7.1 is a histogram of some actual measurements and is provided to give the reader an indication of some "typical" latencies at the process level. The measurements were made on a relatively modern processor and multitasking OS under a modest load and are not inconsistent with similar measurements made with

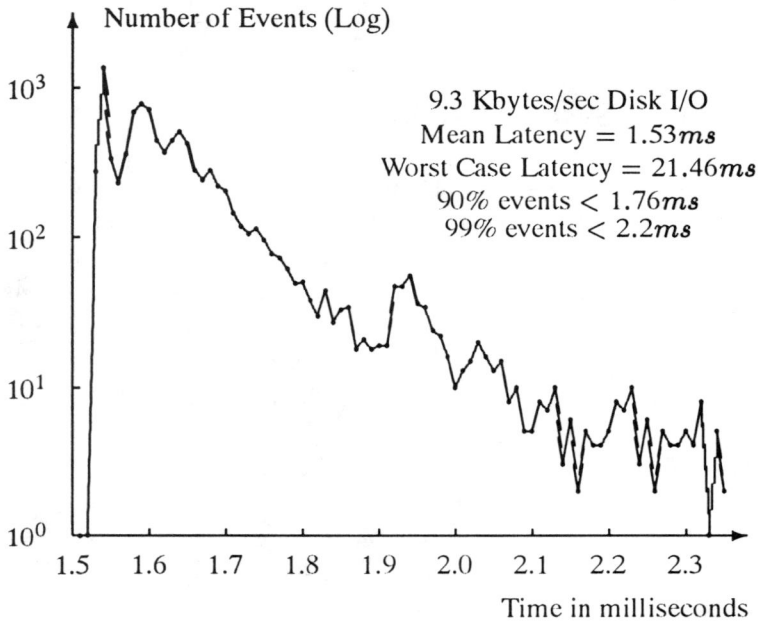

Figure 7.1: Typical OS Process latency histogram

other processors and OS. The measurement does include active disk I/O.

7.4 Buses, Bandwidth and Latency

While not as dramatic as OS latencies, bus latency and bus contention are important considerations when configuring systems and trying to estimate realizable throughput. Here again, be cautious of the conditions that are represented in vendors' claims. All buses are not equal, and published "theoretical bus bandwidths" are not always good indicators.

7.4. BUSES, BANDWIDTH AND LATENCY

7.4.1 Backplane Buses

The goal of this section is to discuss some of the concerns surrounding a few common buses used for data acquisition and control. This discussion is not intended to be a comprehensive study of buses.

VXI and VME

The VXI and VME buses are essentially the same since the VXI bus is an extension of the VME bus, and, as far as data transfers are concerned, they are identical. In its most frequently encountered implementation, the VME bus is a 24-bit wide address and 32-bit wide data bus with a 100ns bus cycle. In *burst* mode it has a theoretical bandwidth of 40Mbyte/second (one 32-bit transfer every 100ns). Supported are 8-bit, 16-bit, and 32-bit transfers. The VME bus is a multi-master asynchronous-handshake bus, where any node on the bus can, in principle, become master and address any other (slave) node on the bus.

A number of factors influence realizable throughput of the VXI or VME bus. An important factor is to recognize that the full 40Mbyte/second theoretical throughput can only be achieved in a dedicated burst mode which may include some undesirable side effects for some applications. The following factors should be considered:

- Width of data transfers between two devices: Some devices only support 8- or 16-bit transfers. Many ADCs generate 12- or 16-bit data which fits nicely into a 16-bit word. The *theoretical* bus transfer rate drops to 20 or 10 Mbytes/second when transferring 16- or 8-bit data, respectively. Some ADC designs pack two successive samples into a single 32-bit transfer, which is both faster and makes more efficient use of the bus.

- The VME bus is an asynchronous handshaked bus: When a bus master addresses a slave device, the data transfer rate is determined by the slowest device. Many VME bus interface implementations limit transfer rates substantially below the theoretical bus bandwidth. Also, inherently slow devices may further slow bus transfer rates when addressed, e.g., slow-access memory. Not only do slow devices take longer when addressed, but they also introduce access latency for intrinsically fast devices by delaying their access to the bus.

- Burst mode: The VME bus supports burst mode where a device arbitrates for the bus and passes the address once, followed by some number of data transfer cycles, with the addressed device performing the auto-increment of the address. The VME specification permits up to 256 byte bursts. A 256 byte burst of 32-bit transfers will block other devices on the bus for a *minimum* of $6.4\mu s$. Depending on the speed of the master and slave as well as whether the transfer is 8-, 16-, or 32-bit, other bus activity could be blocked for a substantially longer time.

CAMAC Dataway

CAMAC is a standard designed specifically for data acquisition and control. The CAMAC backplane bus is known as the Dataway. The Dataway is a single-master synchronous 24-bit-wide bus with a bandwidth of 3Mbytes/second. Provision is made for multiple controllers or masters with a strict-priority hardwired arbitration scheme provided between masters (controllers). Most I/O devices (modules) are slave devices only and must accept or provide up to 24-bits of data within the $1\mu s$ bus cycle, or provide a signal (No-Q) that they are not ready to accept a data transfer.

In CAMAC, throughput rates may be limited by the Dataway bandwidth. However, in most systems the limitation is frequently the ability of the controller to transfer data to computer.

7.4.2 Interconnect Buses

Interconnect buses are a class of buses that are designed to interconnect independently powered chassis or devices to a computer system. SCSI and GPIB are popular examples that are available on a wide variety of platforms. These buses are typically byte or word serial depending on the implementation and support throughputs of 1 Mbyte/second and higher. They include the capability to address multiple devices on the bus and support multiple bus masters.

SCSI

SCSI is an interconnect that was originally developed as an I/O path to disks and tapes. Since most workstations provide an external SCSI port, this interconnect has become widely used for data acquisition as well—especially since many workstations do not provide an external general-purpose I/O bus.

On the surface, SCSI looks like a widely available high-performance I/O bus with throughput well over 1Mbyte/second. For moderate performance data acquisition and control applications, SCSI performs well. SCSI becomes a limitation in the latency area. Disks and tape which drove the SCSI implementation are electromechanical devices that have access times measured in the tens of milliseconds and transfer relatively large blocks of data at rates in megabytes per second. SCSI is well matched to these requirements. Unfortunately, SCSI is not as well suited to transferring small blocks at low latency that might be encountered in a 100 Hz control or data acquisition application. Also, since disks do not tend to post asynchronous events with any frequency or urgency, SCSI does not provide a low-overhead mechanism for Asynchronous Event Notification (AEN).

The other limiting consideration with SCSI is the fact that the host computer interface must be designed to handle a class of devices on the SCSI bus. This is accomplished with a port and class driver approach. The port driver handles the physical SCSI port and there are individual class drivers for disks, tapes, and other devices. Thus, an I/O request must pass, in effect, through two drivers, one *class driver* (e.g., disk, tape, or data acquisition device) and the SCSI *port driver*. This structure adds overhead and latency.

While the above discussion may seem somewhat negative, SCSI nonetheless, does provide a good general-purpose alternative for many applications. By careful design of the data acquisition front-end with adequate buffering and intelligence, overcoming many of the limitations is possible. By applying some of the techniques discussed in Section 7.5, the effects of latency can be minimized for data acquisition applications. By incorporating a local processor at the I/O chassis-level, some of the latency issues in control applications can be overcome. Some of the latency issues caused by contention for the bus can also be controlled by placing the data acquisition devices on a separate bus.

GPIB

The GPIB bus is a widely used interconnect for instrumentation applications, and is available on many host processors. It was designed originally to provide a communications path to stand-alone instruments such as digital voltmeters, frequency generators, etc. which normally accept ASCII commands and generate ASCII data. There is, however, no restriction as to whether the data is ASCII or binary. GPIB throughput is limited to approximately 1 Mbyte per second, and suffers from many of the same latency issues as SCSI. Unlike SCSI, it usually does not require the multiple levels of operating system drivers and thus does not incur this extra latency.

7.5 Buffering and How It Helps Overcome Latency

In general, the basic data acquisition devices, such as ADCs, generate data (converted values) at rates that are based on a hardware clock. The rate may range from sub-microseconds to seconds, but when the "clock ticks," data is sampled, converted, and presented at the output following the conversion time. The data will generally remain valid either until the next clock or the end of the next conversion. In any case, there is a fixed time interval that *something* must respond and move the data to a storage device.

Depending on the architecture of the data acquisition system, this scenario is repeated multiple times, whether it be for a single sample from a single channel; a buffer of data corresponding to all the data from multiple channels at a single time interval; or a buffer covering multiple time intervals. In any case, there is a finite time interval where *something* must respond and move the data before the next block starts. Failure to respond in time means *lost data*.

The simplest solution is to interrupt the processor each time the ADC has a value and have the CPU place the value in a buffer (block of main memory), and when it is full write the buffer to disk or other permanent storage. This technique works until the interval between ADC conversions approaches the worst case latency of the application software in the CPU. From this discussion of latency, this data rate is relatively low. Buffering of data at the hardware level is the technique that permits data rates approaching the I/O bandwidth of the CPU or storage device, whichever comes first.

In modern data acquisition systems, several buffering techniques are used and may occur at multiple levels, including at the ADC module, at the I/O chassis controller level, at the host computer interface level, within the host computers main memory, or within the disk controller. The purpose of buffering is simply to provide a place to store data for a period that is *long* compared to latency associated with accessing the affiliated device.

For example, if the ADC conversion time is short compared with the worst case bus latency that the ADC uses to transfer data, then *the ADC must include sufficient buffering to insure no data is lost*. If the device that initiates the transfer of data from the ADC is a processor, then the amount of buffering must be consistent with the worst case latency of the processor plus I/O bus.

7.5.1 Buffering Techniques

A number of buffering techniques exist. The simplest is double buffering. In this case two buffers are used. When one is filled, a flag or interrupt is posted indicating that the first buffer is full, and the device, e.g., the ADC, begins filling the second buffer while the first is being emptied.

Multibuffering

A generalization of double buffering is *multibuffering*, where a number of buffers N with $N > 2$ of equal size are defined. The I/O device fills them in order, signaling the processor as each is filled. It is the processor's responsibility to insure that each buffer is emptied before the I/O device is ready to fill it again. The advantage of multibuffering is that when dealing with *very long* worst case latencies, but relatively shorter average latencies, the processor can act on the data faster *on the average*.

In the latency example shown in Figure 7.1, the buffer size would have to be large enough to accommodate the 22 ms worst case latency L_{wc} to prevent loss of data for

7.5. BUFFERING AND HOW IT HELPS OVERCOME LATENCY

double buffering. Assuming a data *fill rate* F_{rate} of 100,000 samples/second and an *average* bus throughput T_{avg} of 2,000,000 samples/second, then the minimum double buffer size B_{size} must be such that:

$$B_{size} > F_{rate} \cdot L_{wc} + F_{rate}(B_{siz}/T_{avg})$$

or

$$B_{size} > \left(\frac{F_{rate} \cdot L_{wc}}{1 - F_{rate}/T_{avg}} \right) = 100,000 \times 0.022/(1 - 0.05) = 2316_{samples}$$

The average "age" of the data accessible by the processor would be greater than 11.5 ms. In the case of quadruple buffering, the minimum buffer size would be about one third the double buffer size and the average "age" would drop to somewhat over 3.86 ms. Furthermore the *total* minimum buffer space required would drop from $2 \times 2316 = 4632_{samples}$ to $4 \times 772 = 3088_{samples}$.

Circular Buffering

Circular buffering is a variation on multibuffering, where each of the buffers appear sequentially in memory. The data acquisition device simply treats the multibuffers as a giant single buffer and resets itself to the head of the buffer when it reaches the end. The advantage is a simplification of the hardware with no loss of generality from the software side. In this configuration, a good practice is to provide a flag for each buffer segment so the hardware can set the flag when the buffer is filled; software can clear the flag when it has emptied the data; and the hardware (and software) can detect a buffer over-run condition should the software fail to keep current.

Ping-Pong Buffering

For some configurations, "Ping-Pong" buffering can be advantageous. This is simply a form of double buffering the data associated with each "tick" of the sample data clock. For example, in a multiplexed ADC with a 10 KHz sample rate per channel, when the 10 KHz sample clock "ticks," the ADC presents one side of the Ping-Pong buffer to the I/O bus while it digitizes values into the other side. Since ADCs do not digitize values instantaneously, this technique gives the processor that is collecting and buffering the data a *full clock period* to move the data.

The Ping-Pong technique can be equally effective on a multi-channel Digital-to-Analog (D/A) converter, in that the processor can load the next set of values for all the channels, and then at the next "clock tick," all the updated values become true. Again, the processor has a full clock period to update the values.

FIFO Buffering

First-In-First-Out (FIFO) buffering basically uses a dual-ported memory where data is clocked in one port and clocked out the second port. These memory devices are typically implemented with RAM memory and two address pointers that keep track of where the next input is to be stored and from where the next output is to be taken. Flags are typically provided for *full, half-full, and empty*. FIFOs are particularly useful for buffering internal to hardware and can be effective as buffers in data acquisition.

FIFOs have two limitations. First, they are generally only available in limited depths. The other limitation is that if the data acquisition process ever gets out of synchronization or the FIFO overflows, the only way to recover is to stop the data acquisition process, clear the FIFO and restart. In other buffering techniques, simply skipping over the buffer in error to regain synchronization is sufficient.

7.6 Using Dedicated Processors to Overcome Latency

In some applications, primarily control, a necessary step is to *initiate action* based on current realtime data. Buffering techniques *do not* solve this problem. The required procedure is to access the input data, make some determination based on the data, and generate a response within some fixed known time interval. When this time interval is shorter than the worst case system latency (e.g., the computer OS latency), then a dedicated processor may be a viable solution.

Several approaches are possible. One approach is to use a dedicated processor with a realtime kernel OS. Here the "dedicated" processor can frequently be used for several related realtime tasks. The realtime kernel operating systems are generally more deterministic and exhibit somewhat lower latencies than larger general purpose operating systems, but one must remember that, *they also have significant latencies* and *ultimately are subject to the same phenomena that cause latency in general purpose operating systems*. The primary advantage of the realtime kernels is that the user has better control over the operating system and its behavior.

The ultimate solution to latency is the *truly dedicated processor* where the application is the *sole* process being executed. Typically, these applications have either no operating system, or just enough of an OS to get the application loaded and to provide some very basic services like I/O drivers and possibly some debugging tools.

In some cases (especially where very fast responses are required) the only solution may be hardwired logic. For example, it may be required to trigger an event within microseconds of an analog signal exceeding a specified threshold. This is a relatively simple feature to implement in an ADC. Such a feature could be used to provide a *trigger*

7.6. USING DEDICATED PROCESSORS TO OVERCOME LATENCY

to initiate the storing of the data in memory, or as a signal to initiate a shutdown of a critical process.

Chapter 8

Types of Data Acquisition

Data acquisition can be generally divided into two classes, continuous and transient. The primary distinctions are the length of time over which the data acquisition occurs, and, to some extent, the sampling rate. At very high sampling rates, the required data throughput can exceed the ability of the storage device, such as a hard disk or magnetic tape, to receive the data. Although disk and tape technology continue to improve, and techniques that divide the data stream into multiple streams each onto separate disk drives can be supplied, continuous high-throughput data storage can be quite costly.

In many applications, it is possible to identify when events of interest occur and limit the high-speed data acquisition to a relatively short time interval of interest. This is considered *a transient application*. When a limited number of samples is required, the best approach is to store the data in high speed RAM memory that is associated with the ADC, then use conventional techniques to read the data from the RAM memory at a slower rate. The limitation of this technique is the size and cost of the local RAM memory.

8.1 Continuous

The function of a data acquisition system is to collect and store the digital information in such a way that it is possible later to reconstruct the time history of any signal as well as its time relationship to other signals. In high-performance continuous data acquisition, the digitized information from input devices, such as ADCs, must be collected and buffered. The data may emanate from a small number of channels at relatively high sampling rates or from hundreds to thousands of channels, with each input being scanned at a relatively modest sampling rate. Often, the most important parameter is the *aggregate* sampling rate. This is calculated as the sample rate per channel *times* the number of channels. This is the data rate that must be handled by the buffer and storage device.

When dealing with one or two channels, it is possible to provide a separate set of buffers and disk files for each channel. As the number of channels increase, it is necessary to use a single set of buffers for the entire acquisition data stream. This reduces computer overhead and limits I/O contention. When a single set of buffers is used, the data words from each channel for a particular scan are buffered in sequence. This is then followed by a data block for the next scan. For some applications a further step that includes a "time stamp" data channel may be necessary. Time stamping is particularly useful when the data acquisition takes place over a distributed area where it is not easy to insure correlated clocking of the data acquisition, or where the data must be time-correlated with some external event or process.

In configuring large continuous data acquisition systems, it is important to analyze carefully the throughput and latencies of the data path between the front-end I/O and the storage medium. Buffering must be provided to accommodate the worst case latencies so that information is not lost because of a data overrun in some element in the path.

Some applications require multiple sampling rates. This leads to an additional system consideration. There are basically two methods to handle these multiple rates: assign a separate data stream and set of buffers to each sampling rate, or choose sampling rates that are integral sub-multiples of the highest rate (e.g., 1 KHz, 5 KHz, and 20 KHz) and use a single set of buffers.

In the first case, each data stream is handled separately. Great care must be taken in analyzing contention for any *shared resources*, such as buses, disks, controllers, and the CPU. If contention exists, adequate buffering must be provided for the worst case. The issue of contention usually becomes unmanageable when more than two data streams are used—especially when the application is stretching resources to the limit.

When the sample rates are integrally related, and a single data stream and associated buffers are provided, data words are identified by their position in the buffer. As more data rates are included, the ordering of data in the data stream becomes more complex. However, with proper system design, a single data stream can suffice for these applications. The single data stream is usually the most straightforward type to implement and provides the most robust data acquisition system.

8.2 Transient

Transient data acqusition systems are usually the simplest and most straightforward to implement. In some sense all systems are transient—no one takes data forever. In practice, what separates the transient type of data acquisition from the continuous type is the amount of high-speed memory that can be dedicated to an ADC front-end. Also, at very high sampling rates, where the net throughput exceeds the bandwidth of the disk

8.2. TRANSIENT

and tape storage devices, transient data acquisition techniques must be used. For some applications, even at relatively modest sampling rates, transient-type data acquisition can be used and may be more cost effective and simpler to implement than continuous data acquisition.

For transient data acquisition, the ADC is interfaced directly to a "local" RAM memory that generally is not part of the CPU's memory bank. In the typical transient recorder, when the ADC scanning is started, a segment of the memory is used as a circular buffer to store the digitized data. Upon receipt of a "trigger," a pre-programmed number of "post trigger samples" are stored, and the ADC stops storing data in the memory. At this point, the digitized data is present in the transient recorder memory. This memory can then be read, starting with the oldest data first. Note that the ADC stores information in the circular buffer before the trigger is received. By allocating a portion of that buffer to the time before the trigger *and* after the trigger, pre- and post-event information are stored.

Some transient recorders offer a "multi-hit" capability, allowing multiple triggers, in sequence, to store a block of data associated with each trigger. These recorders operate in a manner similar to that described above, but with two important differences. First, no "pre-trigger" samples are allowed. This insures that the data always starts in the memory from the start of the trigger segment. Second, each "hit" is stored in a successively higher memory segment. Most recorders with multi-hit capability can also be programmed for single-hit operation with the associated pre- and post-trigger functions.

What simplifies the transient recorder application, as far as the data acquisition system is concerned, is that sufficient time is generally available to read the memory before the next set of samples needs to be taken. This avoids many of the throughput and latency issues associated with continuous data acquisition.

Chapter 9

Systems Considerations

9.1 Calibration

The most important function of a data acquisition system is to ensure integrity of the measurements produced. The data must be an accurate representation, to the highest practical degree, of the physical values that are being measured. If, prior to a test, a data acquisition channel is not ready to produce "good" data, this condition must be detected and reported, so that corrective action can be taken. The problem may be caused by such items as a faulty transducer, input wiring problems or a failed component in the signal path.

Calibration can help accomplish the goal of assuring reliable data by verifying the integrity of the signal path. As discussed earlier, calibration also helps to reduce system costs by placing the burden of absolute accuracy on relatively few components.

The transfer functions of many transducers, i.e., the relationship between the physical values being measured and the output voltages or currents, cannot be included in the calibration cycle of a data acquisition system. This is because the physical value—such as temperature—must be changed to perform the desired measurements. Calibration factors for such sensors are derived from manufacturers' data sheets, standard tables, or the characteristics of individual transducers as determined by a calibration laboratory. These values are then entered into the computer. Some transducer subsystems, such as multiplexed pressure sensors, actually are calibrated while attached to the system—by injecting precision pressures into the subsystem. Many data acquisition systems are able to detect open or short-circuited transducers or wiring. For example, an "open" thermocouple can be detected by passing a current through the thermocouple wiring and using the data acquisition system to measure the resulting voltage. This arrangement provides a measure of the source resistance to the thermocouple circuit.

9.1.1 Voltage Injection

With the voltage-injection method, a known voltage is inserted into the input of a channel and the digitized response from the ADC is monitored. If three voltage points are chosen, e.g., near + and - full scale and zero volts, the transfer function of the channel can be easily computed. In addition, if this transfer function falls outside prescribed limits, a warning can be generated.

9.1.2 Bridge Transducer Verification

For bridge-type transducers, shunt calibration generally is used. This type of calibration is performed by taking at least two sets of readings, one set with and one set without a shunt resistor connected across one leg of the bridge. It is recommended that a number of readings be taken in each state and the results averaged to minimize the effects of noise. In addition, a check should be incorporated to measure the range of a set of values, thus providing an indication of the noise level and perhaps identifying installation or system problems. If the data acquisition system uses a programmable relay to connect the shunt resistor, this calibration can be performed automatically.

If bridge balance is provided, an additional level of transducer verification may be performed. Assuming that the bridge balance circuitry operates by injecting current into the bridge, the resulting voltage shift provides a measure of the bridge resistance.

Ideally, the excitation voltage should be sensed at the bridge. This allows the actual voltage at the transducer to be measured, and the effects of wiring resistance nullified. The signal conditioner should be able to measure the excitation voltage to calculate the transfer function of the bridge as well as to verify that the correct excitation voltage is present.

While the approximate resistance of bridge-type transducers can be determined, the actual transfer function of that type of a transducer, such as a strain gage, can only be determined from manufacturer's data or by changing the physical value—strain, in this case.

9.1.3 Reference Voltage

All calibration voltages must be referenced to some highly accurate standard, such as provided by NIST (National Institute for Science and Technology, formerly the National Bureau of Standards—NBS). This requires periodic calibration of the data acquisition system components, so that any inaccuracies can be corrected. Instead of adjusting the gain and offset for all channels, it is desirable to reference all calibration measurements back to one point in the system. Thus, if this system reference is used as the basis for all calibration, only that component need be calibrated periodically. The data acquisition

9.1. CALIBRATION

system should provide a means for busing this system reference voltage to the voltage injection relays for each signal conditioning channel.

9.1.4 Calibration Archiving

Often, it is not sufficient to simply archive the data from a test. Responding to regulatory requirements, customer specifications or a desire for more data security, it may be necessary to archive the calibration information along with this data. This approach allows the accuracy of the data to be verified at a future date. With the calibration procedures discussed here, the additional archiving can be performed quite simply. The data acquisition system can produce a complete calibration record, taken just prior to the test. The "raw" uncalibrated data may also be archived.

> **One of the leading airbag manufacturers** *demonstrated the importance of built-in diagnostics in front-end modules. They wanted their test system to notify them immediately if a transducer failed or if a problem with the sensor wiring occurred. With built-in diagnostics, designed by KineticSystems Corporation, these types of error conditions (excitation alarms, open/short conditions, etc.) can be reported immediately and also logged into the datafile.*

Chapter 10

Software Considerations

Software support for data acquisition and control applications is increasingly important. In the past, much of the data acquisition software has been in the form of either *support routines* or dedicated turnkey packages. There are a number of software packages that have been developed and supported by process control system vendors. Typically, these packages only support the various vendors' hardware. Until recently, few general-purpose data acquisition system software packages have been available. Most of the data acquisition software packages had been dedicated to highly specialized applications.

Over the past few years, with the advent of graphical workstations, a number of Graphical User Interface (GUI)-based, general-purpose data acquisition software packages have become available. An increased priority among users is the specification of open-systems software and hardware *based on recognized standards*. Software standards, such as POSIX, increase the portability of an application; hardware standards, such as CAMAC and VXIbus, help to ensure a source of products from a number of vendors. Truly distributed systems using Ethernet or some other networking approach, are a key to many larger applications. The configuration and use of larger systems are simplified by including data acquisition software that uses a database as its core. This chapter will explore some of the issues and features associated with data acquisition software. It is not intended as a comprehensive survey.

10.1 Data Acquisition Software Features

In evaluating the features of various data acquisition software, it is important to note that a system *must* address several distinct phases of operation:

Setup or Configuration: During this phase, the data acquisition system hardware and software must be configured for a particular application or a test run. This includes defining which channels are to be used, setting channel gains, selecting filters, determining sampling rates, etc. This phase may also involve determination of which signal conditioning and digitizing modules are physically present and operational.

Calibration and System Checkout: During this phase, the entire system should be verified for proper operation. In many cases, this involves a channel-by-channel calibration to establish the transfer characteristics of each channel. For completeness, it is very desirable that the system archives the results of the system calibration. This information can be invaluable, should later analysis of the data suggest that there may have been a malfunction of the equipment or the sensors. To minimize the effects of system drift, calibration should be performed as near to the system initialization as practical.

System Initialization: During this phase, all system components are prepared for the data acquisition phase—just prior to acquiring the data.

Data Acquisition: During this phase, the system collects data, archives selected data to storage, and, typically, provides some level of monitoring by the operator and/or users. Realtime monitoring generally is used to verify the *quality* of the data being acquired or to provide a "quick look" at the process being monitored. In some applications, the data acquisition software provides monitoring of some of the data for out-of-limit conditions. Further, the software may transmit alarms to the operator, and, in some cases, may take the necessary control action to automatically bring the device being monitored into a safe operating range, or may halt the test. Some data acquisition system applications also include closed-loop control operations.

Post Acquisition Analysis: During this phase, it may be necessary for the system to provide various standard reports, including summaries of the data that has been acquired. In some cases, this phase supports a complete analysis of the acquired data.

It is the function of the data acquisition software to facilitate these phases of operation.

With a larger system, it becomes critical that the software automates these operating phases to the extent possible, and that the hardware provides the necessary underlying functionality. A user may have a large system with thousands of channels that include many identical front-end modules which do not support geographic addressing—the

10.1. DATA ACQUISITION SOFTWARE FEATURES

logical addresses must be set by switches on the modules. Perhaps while preparing for a test, the slot positions of two otherwise identical modules are transposed. Since the software *can only* associate the data with the module address, the resulting data would be associated with the "wrong" transducers. If the modules support geographic addressing, the software automatically conforms to the new configuration.

10.1.1 Large versus Small System Considerations

As mentioned above, larger applications with high numbers of input channels require significantly more software functionality than smaller applications. For a small application, the system generally is controlled by one person or a very small group of people, and the number of input channels are limited. A user of a small data acquisition system is more likely to be aware of any significant changes made to the system than a user of a large system.

In larger systems, it is much more difficult to insure manually that the physical system is properly configured and matches the software configuration. The large number of channels may preclude manual checking because of the human resources required. Also, the ultimate consumer(s) of the data may have little knowledge of the detailed system configuration and must be able to access data at the more abstract level of a *logical name* which is associated with a sensor. In addition, such a user will likely want to access the data expressed in engineering units corrected for calibration. In larger systems, it is increasingly important that the software be capable of assuming this higher level of functionality.

Features that are important to larger systems include:

- Automated calibration procedures with the ability to archive the calibration information as well as to associate it with the data.

- Automated system checkout. This should include the ability to:
  - Verify that the *proper modules* with the *proper options* are located in the desired chassis slots, since users wire signals to physical locations—not to *logical* addresses).
  - Verify that the hardware is operating properly. For analog modules, this often involves the injection of a known analog signal at the input and checking for an expected result.

- The software should provide a facility to assign logical names to signals and permit the retrieval of data by logical name.

- The software should support multiple "test plans" or configurations. Once an operator has generated a *test plan* and verified it, the plan should be saved and easily recalled by name.

- The software should provide for report generation to facilitate the creation of summary reports for a given data acquisition run.

Chapter 11

System Considerations and VXI

One of the short-comings of many product offerings is the lack of an over-all systems approach. While individual system components may exhibit excellent characteristics, when brought together to solve a specific application, deficiencies may become apparent. This chapter discusses some of the system-level issues and illustrates some of the features of the VXI standard which address these issues.

11.1 VXI Bus Features

The VXI C-size specification, while it uses the same bus specification as VME, does provide some significant system level enhancements over VME. These include the definition of 8 TTL and 2 ECL trigger lines, a 10MHz ECL clock for system timing, module ID lines for geographic addressing and module identification, an analog summing bus, and the 12-bit wide daisy-chained local bus. These features of the VXI standard provide a solid basis for a more comprehensive systems level approach.

11.1.1 VXI Local Bus for Inter-module Communications

The VXI Local Bus provides 12 user defined lines on the backplane between the right row of pins on one module and the left row of pins on the adjacent module to the right. These lines are suitably shielded for high level analog or digital signals. For analog signals, these lines provide an elegant solution to the issue of getting signals from signal conditioning modules to the analog-to-digital converter (ADC). For digital signals these lines provide a simple way to implement private data paths between system components that follow the natural data flow. For example, data between an ADC and a Digital Signal Processor (DSP) and/or digital-to-analog converter (DAC).

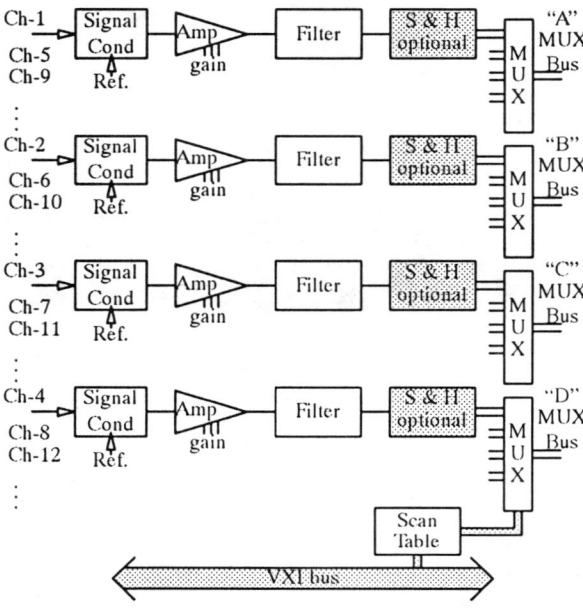

Figure 11.1: Typical signal conditioning module

11.1.2 Local Bus for Routing Analog Signals

Traditional ADC design places the multiplexers and ADC in a single module, and places the signal conditioning either in a separate module or, in some cases, mezzanine cards in the ADC. When the signal conditioning is incorporated in the ADC, usually the number of channels and/or signal conditioning options are limited. When separate signal conditioning modules are used, a wiring maze results. In a departure from traditional ADC design, a more effective approach is to move the multiplexing and simultaneous sample-and-hold functions out of the ADC module into the individual signal conditioning modules. An example of such an approach is illustrated in Figure 11.1. For large signal count applications this has several distinct advantages:

- It eliminates the wiring maze between signal conditioners and the ADC.

- It improves over-all system reliability, since all interconnections between the ADC and Signal conditioning is via the backplane—no messy cables that can develop intermittents.

11.1. VXI BUS FEATURES

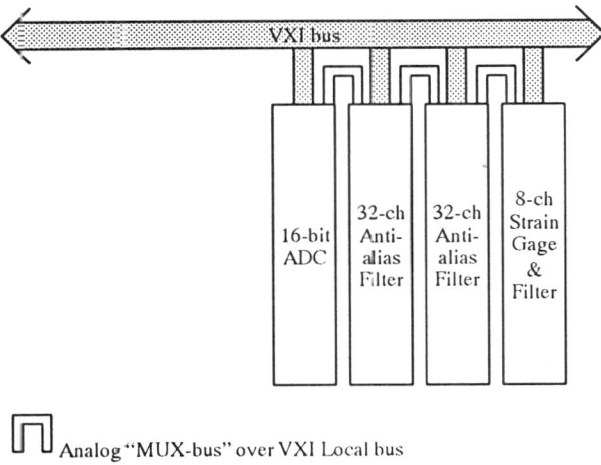

Figure 11.2: Typical MUX-bus configuration

- It reduces discrepancies between actual physical wiring and software configuration parameters, since routing between the signal conditioning and ADC is now completely configured by software.

- It simplifies configurations with different signal conditioning needs, since signal conditioning modules can be chosen to match application requirements and configured into a single subsystem.

- It is easily adapted to a wide variety of applications from low-channel count, high sampling rate to high-channel count, moderate sampling rate applications.

With this architecture, the multiplexed analog bus or "MUX-bus" extends to the right of the ADC and is propagated from slot to slot by signal conditioning modules. Figure 11.2 illustrates the relationship between signal conditioning front-end modules and the ADC using the MUX-bus concept. The MUX-bus architecture supports an arbitrary number of input channels per ADC distributed over an appropriate number of signal conditioning modules. A *typical* implementation might support 256 channels as a trade-off between cost per channel, support hardware, sampling rate, and what will physically fit in a VXI chassis. The number of channels supported by a single signal conditioning module can range from a minimum of 4 channels to the full implementation limit in 4-channel increments.

To achieve optimum throughput and accuracy, MUX-bus employs a 4-phase differential channel multiplex where three channels are in different stages of settling, while one channel is sampled and converted. In addition MUX-bus supports a common precision reference voltage supplied by the ADC as well as timing and control to synchronize the multiplexors across multiple signal conditioning modules.

11.1.3 Local Bus as a Private Digital Path

In many applications there is frequently a requirement to capture analog and digital data, and either process it in some manner in realtime, or to process it and generate some form of output. The former includes such operations as performing FFT's, digital filtering, and signal averaging. The latter includes high performance control loops and the generation of some waveform or analog or digital output which is dependent on input data. These requirements often dictate that the operation be synchronous with either some clock or externally supplied signal.

The VXI Local Bus again provides a convenient mechanism to implement a private digital bus or "Digi-bus" for moving synchronous data between modular elements of the system. By using a private synchronous bus, as opposed to the main VXI backplane bus, issues regarding bus contention and latency are avoided. For example, the bus interface can be implemented on an ADC or Digital Input as a "data source," while modules such as a DAC and digital output modules are implemented as "data sinks". Processor modules such as a DSP may act both as sources for processed data and sinks for raw input data. Figure 11.3 illustrates how Digi-bus could be used to interconnect an ADC, Digital Input, DSP, and DAC modules.

Selection of a bus protocol is important. The issue of multiple sources and sinks as well as device addressing must be defined. As protocol complexity and general purpose utility grows, effective bandwidth drops and bus latency increases. Since this is a dedicated private bus, an effective solution is a simple synchronous protocol based on the concept of a "frame" of data where each data source has pre-assigned locations within the frame to place data, and data sinks can extract data from *known* locations within the frame. One data source acts as the "master" providing any necessary timing signals. The entire data frame is available to all modules connected to the Digi-bus segment. Note that since the Local bus is propagated from slot to slot, multiple Digi-bus and MUX-bus segments may co-exist in a common VXI chassis. Data frames are generated synchronously with the "data acquisition" or "sample" clock.

Since a particular "sink" module may only require selected data from a frame, or in some cases, only every Nth sample, a data selection scheme must be provided. Two simple selection schemes can be easily implemented. One is simply a bit map of the frame that selects data based on position within the frame, and the second is to only pass

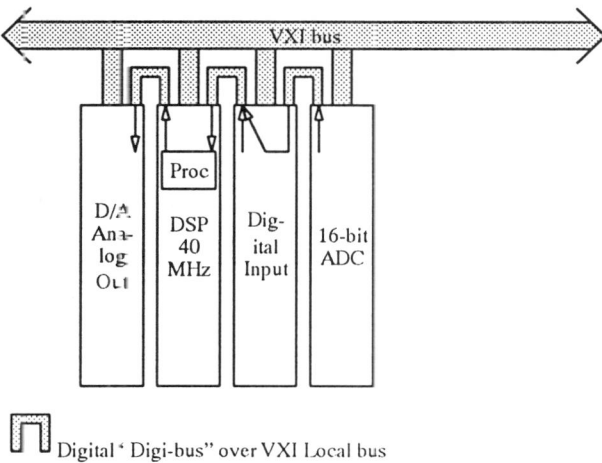

Figure 11.3: Typical Digi-bus configuration

every Nth frame. For example, a 16-channel DAC might select the 2nd, 9th, 10th, and 23rd channel from every 10th frame to route to analog output channels 1-4.

11.1.4 Data Flow

A less obvious issue is selection of a direction for data flow. The key element here is the VXI Slot-0 controller which is defined by the standard to be in the far left slot. Since the Slot-0 is either a computer in its own right (smart controller) or is connected to a computer, it is desirable to not preclude the possibility of passing digital Local bus data directly to the Slot-0. This means that the "Digi-bus" must exist to the left. Since an ADC typically might implement both "Digi-bus" and "MUX-bus", MUX-bus must exist on the right side of the ADC with the signal conditioners immediately to the right of the ADC. A typical configuration is illustrated in Figure 11.4.

11.2 Trigger Lines for Time Synchronization

For example, in the VXI specification, eight of the pins on the P2 connector of the VXI bus are defined as TTL trigger lines. These open collector lines provide a wired-OR function that is ideally suited to their use in communicating event information between modules.

A useful concept in using these lines is one of "Event Sources" and "Event Sinks". Thus any one of several modules may be capable of generating an "Event" on a specific

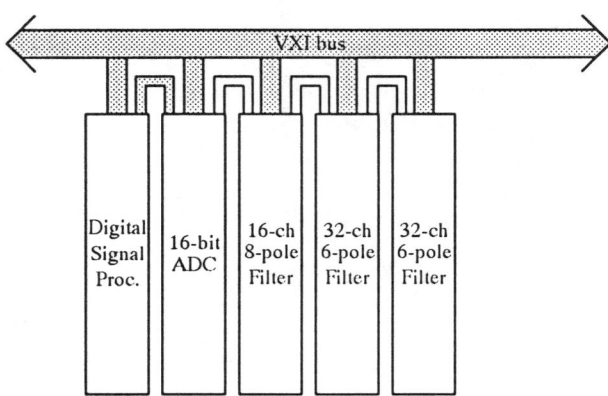

Figure 11.4: Typical system configuration illustrating MUX-bus and Digi-bus

trigger line, and one or more modules may be programmed to respond to this "Event", including the module which is the source of the event. Examples of "Event Sources" would be sample clocks, limit checking and front panel external trigger inputs. Examples of "Event Sinks" would be transient capture triggers to initiate the capture and local storage of a data segment, and sample clocks to synchronize sampling of input signals or outputting of DAC or digital data.

Consider the case where a number of transient capture ADC modules are required to trigger simultaneously. The occurrence of a limit condition on any of the modules can cause a trigger line to be asserted and a transient capture initiated across all modules. Meanwhile, another trigger line has been programmed as the source of the sample clock. This sample clock is used by the Signal Conditioning modules to strobe their sample-and-holds into hold mode and by the ADC to initiate conversion. Multiple sample rates can be handled quite easily.

11.3 Overall Architecture

The overall architecture is illustrated in Figure 11.4. Starting at the right, the Signal Conditioning modules provide the necessary signal conditioning, gain, filtering, and multiplexing. By using RAM lookup tables, the signal conditioning modules can be configured to sample selected channels in a suitable sequence. For applications requiring simultaneous sampling, sample and holds can be included at the signal conditioning level. Trigger lines can be used to synchronize sampling across multiple modules. A common

11.3. OVERALL ARCHITECTURE

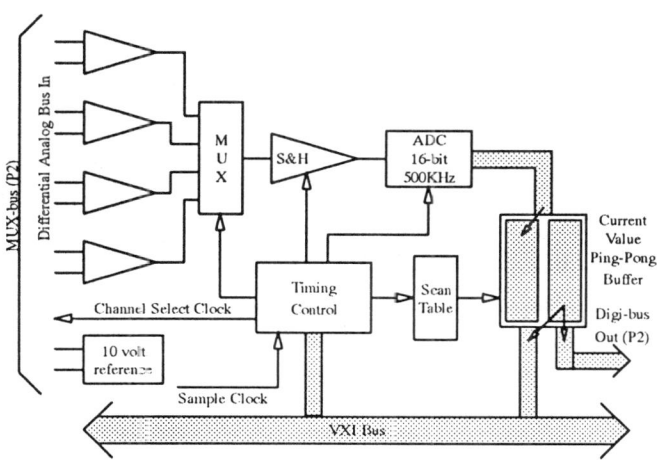

Figure 11.5: Typical ADC configuration

precision reference from the ADC, a local reference, or an external calibration signal can be used for calibration *and* end-to-end system checkout. A *typical* signal conditioning module is illustrated in Figure 11.1. Multiple signal conditioning modules can be mixed and matched to meet a particular requirement.

The ADC provides the analog-to-digital conversion function. Since signal multiplexing and simultaneous sampling are handled in the signal conditioning, there is now space on the ADC to implement other desirable functions. These include a large multi-buffer memory, ping-pong current value registers, Digi-bus port, per channel limit checking, and a precision reference. Refer to Figure 11.5.

The multi-buffer memory provides critical buffering between the Slot-0 processor and the ADC to prevent data loss due to processor latency when switching buffers and processing data. The multi-buffer memory can also provide a "transient capture" function, where a fixed block of data is captured in memory based on the occurrence of a trigger event. Possible trigger events might include an external trigger or one or more of some selected analog inputs exceeding a programmed threshold.

Ping-pong registers are used to hold the data from the previous complete data scan for all channels. The set of registers accessible from the VXI bus are loaded synchronously with the start of a scan and are valid until the start of the next scan. This technique provides the *full* interval between scans for data access.

The Digi-bus provides a private bus to stream data to processing elements such as a DSP which can perform such functions as FFT, Digital Filtering, Signal Averaging, or other desired functions. Trigger lines provide such information as the start of a transient and start of an ADC scan.

The ADC includes hardware to allow limit checking on a per channel basis. This feature permits the application to establish either upper or lower threshold checks on an individual channel basis. When one or more of the channels exceed the individually programmed thresholds, the ADC will generate a trigger or interrupt. This feature provides the application with a very high speed threshold detection that can be used to trigger data capture or signal other hardware to take action.

The DSP provides a Digi-bus sink for input data and a Digi-bus source for processed data. For data acquisition applications, the DSP can provide such functions as FFT of in-coming data, digital filtering, and signal averaging. For control applications, the DSP can provide processed data for generating analog or digital outputs that are synchronized with data input.

11.4 System Calibration and Diagnostics

In any system, and particularly larger data acquisition systems, it is very desirable to be able to verify that the system is functioning properly and to verify the calibration of analog I/O. It is generally possible to do some level of checkout of digital system components through exercising the hardware under software control. However, to check out analog channels to any degree requires that a series of known analog signals be injected into each channel of the system. This can be accomplished either by hand or by designing into the analog front-end circuitry the capability to switch known calibration signals into the input under software control. Switching of calibration signals by hand can be very time consuming and is subject to errors.

For these reasons, it is desirable to choose an implementation with full end-to-end calibration features. To accomplish this goal with the architecture described, it is necessary to be able to inject a known signal at the input to the signal conditioning. This is accomplished by a front-end MUX on the signal conditioning module that can switch between the input signal, a programmable active attenuator, and an external user-supplied front-panel calibration signal (see Figure 11.6). The reference for the attenuator can be selected between an internal lower precision "local" reference and a precision reference supplied via MUX-bus by the ADC. This approach permits stand alone use of the signal conditioning, and distributes the cost of the precision reference over all the signal conditioning modules tied to an ADC.

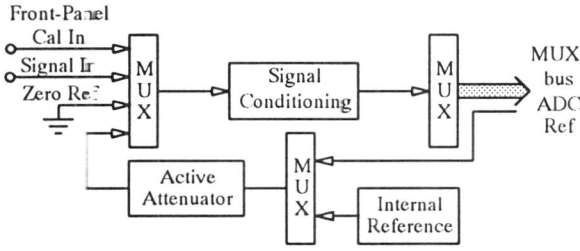

Figure 11.6: Typical calibration configuration

11.5 System Configuration Validation

One of the issues, particularly in larger systems, is the ability to insure that the proper modules with the proper options are installed and configured in the system. This is especially a concern where it may not be convenient or feasible to accurately validate the *current* system configuration due to system size or physical access to areas in a distributed system. This can be particularly frustrating when modules of the same type or model number can be configured with a number of different internal options. Unfortunately, this is increasingly common practice due to the high densities that can be achieved today and the relatively large card size of VXI. A related issue is the ability to identify and trace the history of specific modules within a larger system.

The VXI standard requires, or in some cases suggests, that certain register conventions be followed. These register conventions partially address the issues raised above. Standardized registers include a unique manufacturer ID that is assigned by the VXI consortium, a "device type" or model identifier, a serial number, and a *version* number or revision number for hardware and firmware.

An extension of this concept is to provide a module option identifier, as well as some amount of user writable EEPROM. The EEPROM provides the user with the ability to record any option, calibration, or other module-specific information that may be important to system operation and/or maintenance in non-volatile memory.

Since these registers are accessible by software, it is possible to develop software to verify system configuration at startup, as well as to track modules for maintenance purposes.

Appendix A

VXIbus, an Open Instrumentation Standard

VXIbus is a modular standard that was originally developed to replace "rack-and-stack" instruments. VXIbus is the acronym for VMEbus eXtensions for Instrumentation. The VXIbus specification was created by a consortium of major instrumentation manufacturers. The goal of the VXIbus standard is to define a technically sound instrumentation specification based on VMEbus that is open to all manufactures and is compatible with present industry standards.

The basic relationship between VME and VXI is shown here:

- VXI uses the VME bus protocol for data transfer between modules.

- VXI uses the identical backplane connector pinout as VME for the P1 (top) connector and for the center row of the P2 connector. The two outer rows of the P2 connector are undefined in VME.

- VXI adds many specifications—including the signal definition of the outer rows of P2 and all of P3, if used—to those provided by VME.

VXI provides numerous enhancements to VME when applied to instrumentation. The major improvements are:

- VXI provides larger card options than VME to allow room for sophisticated analog modules and to isolate critical signals from the digital backplane.

- VXI includes mandatory analog power supply voltages.

- VXI provides for shields on C- and D-size modules.

- VXI specifies clock and trigger lines, geographic addressing, an analog summing bus, and a local bus on connector P2—and P3, if used.

- VXI specifies important characteristics, such as shielding and cooling, of the I/O chassis. These chassis are called mainframes.

A.1 VXI Module Sizes

There are three VXI module sizes, defined as B, C and D. The B-size module has the same dimensions as a B-size VME module, with all three rows of contacts on the P2 (bottom) connector fully defined for the VXI module. The relative sizes of the three module types are shown in Figure A.1. For most instrumentation, the clear choice is the C-size module. This configuration provides sufficient board space for sophisticated instrumentation and is deep enough to allow for physical isolation of low-level analog signals from the fast-risetime signals associated with the digital backplane. D-size modules also can be used for very sophisticated circuitry. However, the extremely large board area of this module makes an extremely high-cost building block.

VME and VXI modules can be mixed in the same chassis. However, care must be taken to prevent severe damage to the modules. The outer rows of the P2 connectors in a VXI mainframe include +5 volt, -2 volt, -5.2 volt and ±24 volt power pins. Many VME modules include manufacturer-defined intermodule buses on these connector rows. Therefore, inserting such a VME module directly into the P2 connector will destroy any TTL logic connected to these power pins. These ICs may then short circuit to other P2 contacts and damage logic in the VXI modules contained in the mainframe. Adapters are available from several manufacturers to allow VME B-size modules to be inserted in VXI C-size mainframes. These units "split" rows A and C on the P2 connector to prevent the connection of incompatible signals. The adapters also logically buffer the bus lines so that unacceptably long unterminated signal "stubs" are eliminated.

A.2 The VXI Mainframe

The powered chassis that houses VXI modules is called a mainframe. Since B-size modules do not include shields, a full-size mainframe for these modules contains slots for 20 modules, while C- and D-size mainframes can accommodate 13 modules. On these chassis the module positions are designated, from left to right, slots 0 through 12. The VXI specification covers important mainframe considerations, such as cooling, power supply noise, electromagnetic compatibility (EMC) and backplane construction. VXI can be used for high frequency applications into the gigahertz range. The specification

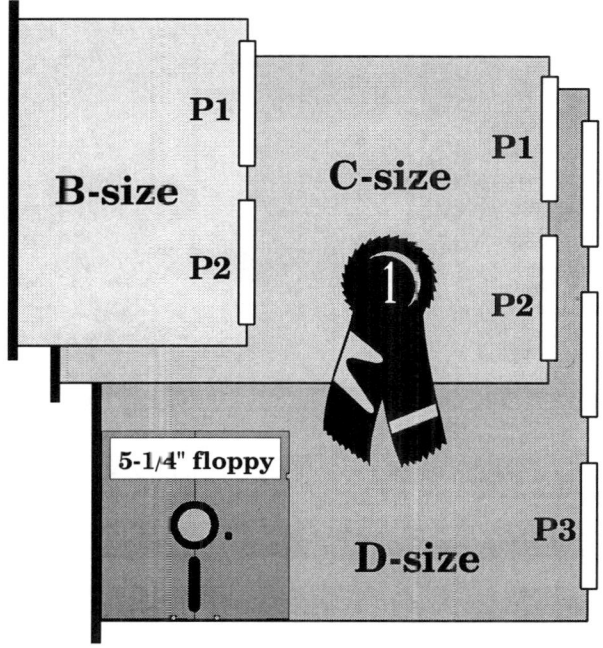

Figure A.1: Relative sizes of VXI module options

allows mainframes to include optional shields between the modules for these applications. More consistent cooling is provided by mainframes that include fans that "push" the air through the modules and that include air baffles at each slot to provide sufficient air flow through the modules in a partially filled mainframe. Careful choice of a VXI mainframe is extremely important for successful system operation.

A.3 The Slot-0 Controller

Since VME—and, by reference, VXI—is a multiprocessor bus, any slot in a VXI mainframe can act as bus master. However, Slot 0, the leftmost slot in the chassis, includes some unique features and backplane wiring. The VXI mainframe "host" is usually an embedded computer that is part of the Slot-0 Controller or an external computer connected

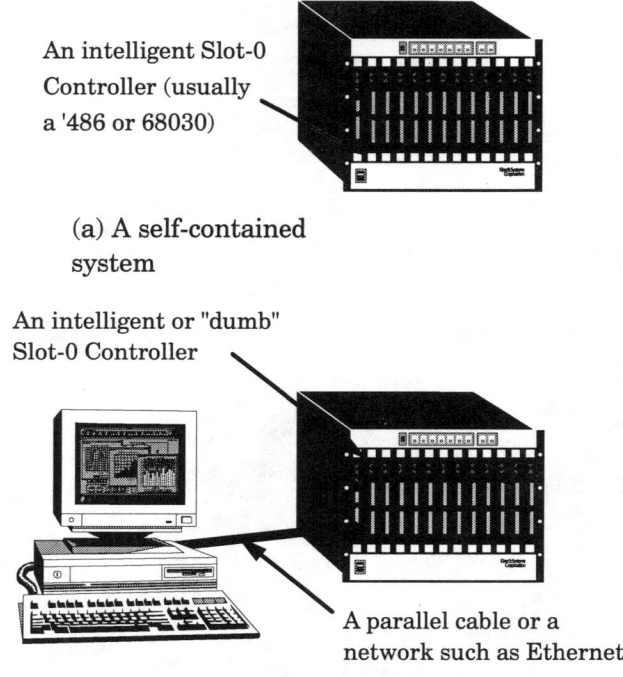

Figure A.2: VXI host computer options

by a bus connector on the front of the controller module. These options are shown in Figure A.2. Popular embedded controllers include 486 PC-compatible and 68030-based processors. Connection to an external computer can be made by use of the MXI parallel bus specified by VXI, Ethernet (with an intelligent processor), IEEE-488 or other interconnection bus.

A Slot-0 Controller provides two primary resources via the P2 connector to the VXI modules, a 10 megahertz common clock and module ID lines. These functions are implemented for the B, C and D form factor. D-size controllers also provide a 100 megahertz common clock on the P3 connecter.

A.4 Using the Module ID Lines

Twelve module ID lines (called MODID) emanate from Slot 0 to the other twelve slots in a 13-slot VXI mainframe. These lines are driven by the Slot-0 Controller and are used for geographic addressing in the following manner:

A.5. REGISTER-BASED VS. MESSAGE-BASED MODULES

- A logical address switch is provided on each VXI module. This switch is used to set the logical base address for the module—the location of a block in memory space for the module's I/O registers.

- If the module address switch is set to "all ones," the module uses its slot position (geographical address) for autoconfiguration of its logical address.

- The resource manager software contains information regarding the physical location of each module and its logical base address.

- During autoconfiguration, the system software interrogates each module via its MODID line to ensure that the proper module is located in each slot. The software can also determine if a module is not present.

- Again, by geographic addressing, the software transfers data to a register that sets the logical base address for that module.

- This process is continued for all modules that support geographic addressing.

While all Slot-0 Controllers must support geographic addressing (MODID), this is an optional feature for all other module types. Particularly for data acquisition and control applications, it is extremely important that the modules support geographic addressing. The primary reasons are given here:

- Incorrect setting of the address registers on modules can lead to severe problems and even equipment damage—if a module controls the wrong I/O points.

- On systems that contain more than one I/O module of a given type, it is fairly easy to switch the positions of two modules after removing them from the mainframe. When the modules are re-cabled, they will be connected to the wrong I/O points if module ID setup is not used.

- Autoconfiguration greatly simplifies system setup, and the software can check that the modules are located in the desired slots.

A.5 Register-Based vs. Message-Based Modules

VXI specifies that modules can communicate over the backplane by register-based or message-based protocol. With register-based protocol, the communication is via an 8-, 16- or 32-bit parallel path directly to I/O registers within the modules. With message-based protocol, an ASCII interpreter is included on each module, and the binary representations

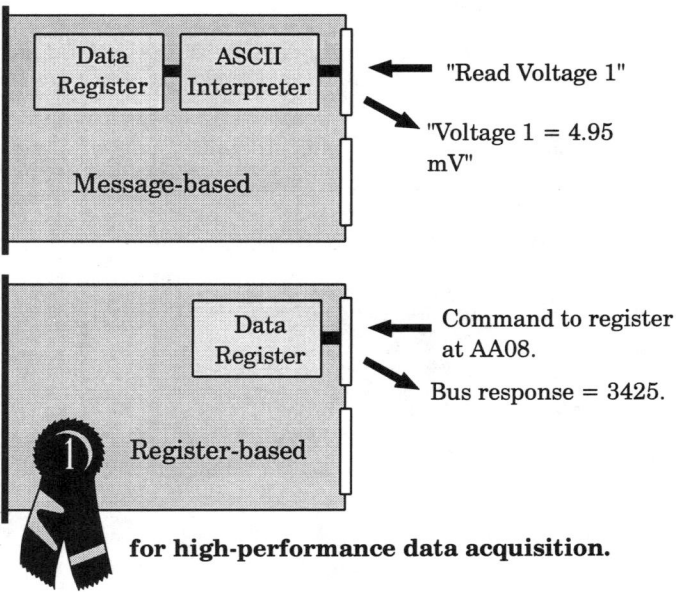

Figure A.3: Message-based and register-based VXI options

of ASCII characters are transmitted over the backplane. These communication options are shown in Figure A.3. The advantage of message-based protocol is that English-like commands and responses can be used. Also, if a consistent lexicon is used, such as the Standard Commands for Programmable Instrumentation (SCPI) specified by the consortium, an instrument from another manufacturer, ideally, could be substituted with little or no change in the software.

High-performance data acquisition and control modules are usually register-based for the following reasons:

- The I/O throughput of a register-based module may exceed 1,000 times that of a message-based module, even if they both are performing the same functions, due to the extensive time needed to parse the word-serial messages in the message-based module.

- Most high-performance data acquisition and control modules contain a number of unique functions, obviating the ability to use the same messages for other brands of hardware without software changes.

A.5. REGISTER-BASED VS. MESSAGE-BASED MODULES

- The tendency is to consider a software driver to be "par" of a VXI module. A properly designed software driver for a register-based module can make the interface to an application program easier, while maintaining high I/O throughput.

Bibliography

[1] J. C. Candy and G. C. Temes, *Oversampling Delta-Sigma Data Converters*, IEEE Press, New York, New York, 1992.

[2] R. T. Cleary, "An Ultrahigh Speed CAMAC Interface for a Large Flight Simulator System," *IEEE Transactions on Aerospace and Electronic Systems*, Vol. AES-22, No. 5, pp. 618-627, 1986.

[3] R. T. Cleary, "The IEEE-583 Bus—CAMAC, A versatile interface standard," *BUSCON-86 Conference*, 1986.

[4] Donald Lewine, *POSIX Programmers Guide*, O'Reilly & Associates, Sebastopol, California, 1991.

[5] Ralph Morrison, *Grounding and Shielding Techniques in Instrumentation*, John Wiley & Son, 1986.

[6] A. V. Oppenheim and R. W. Schafer, *Discrete-Time Signal Processing*, Prentice Hall, Englewood Cliffs, New Jersey, 1989.

[7] S. Park, *Principles of Sigma-Delta Modulation for Analog-to-Digital Converters*, APR8/D, Motorola, Phoenix, Arizona, 1990.

[8] R. W. Steer, "Antialiasing filters reduce errors in A/D converters," *EDN*, pp. 171-186, March 1989.

[9] R. W. Steer, "Eliminating Alias Errors in DA and Logging Systems," *Sensors*, June 1990.

Index

A

accuracy, 45
    specification, 30
ADC, 43
    flash converters, 44
    hybrid converters, 44
    integrating, 43
    quantization noise, 38
    scanning, 46
    sigma-delta, 44
    signal to noise, 38
    successive approximation, 43
    types, 43
aliasing, 9–10
analog
    front-end, 13
    specifications, 27
analog-to-digital Converters, 43
antialias filter
    considerations, 34
    cost considerations, 35
    system cost considerations, 34
antialias filtering, 11, 33
auto-ranging, 25

B

bandwidth, 9
    amplifier, 23

Bessel filters, 35
bridge
    balance, 32
    balance calibration, 64
    completion, 26
    conditioning, 25
    excitation, 25
    shunt calibration, 26, 64
    transducer verification, 64
buffering, 47, 53
    circular, 55
    double, 54
    FIFO, 56
    multi-, 54
    ping-pong, 55
    techniques, 54
bus
    bandwidth, 50
    GPIB, 53
    interconnect, 52
    latency, 50
    SCSI, 52
    VME, 51
    VXI, 51
buses, 51
Butterworth filters, 36

C

calibration, 24, 63

archiving, 65
bridge, 64
reference voltage, 64
shunt, 26, 30
CAMAC, 5
    Dataway, 52
Cauer filters, 36
Chebyshev filters, 36
circular buffering, 55
CMRR, 15
    specification, 31
coaxial cable inputs, 22
common mode, 15
    rejection, 15
continuous data acquisition, 59
continuous filters, 36

D

data
    errors, 2
    quality, 2
    sampling, 9
data acquisition
    continuous, 59
    software features, 67
    transient, 60
    types, 59
Dataway, CAMAC, 52
dedicated processors, 56
delta-sigma converters, 44
differential
    input, 13
digital filtering, 40
double buffering, 54

E

elliptic filters, 36
errors
    in data, 2

F

FIFO buffering, 56
filter implementations, 36
    continuous, 36
    switched capacitor, 37
filter types, 35
    Bessel, 35
    Butterworth, 36
    Cauer, 36
    Chebyshev, 36
    elliptic, 36
filtering, 33
    antialias, 11
    digital, 40
flash converters, 44
frequency
    Nyquist, 9

G

gain, 24
    specifications, 29
    stability, 30
geographic addressing
    VXI, 85
GPIB, 5
GPIB bus, 53
ground loops, 17–18
grounding, 13, 18
    configurations, 18

I

IEEE
    488, 5
    583, 5
input
    coaxial cable, 22
    differential, 13
    isolation, 16
    single-ended, 13
    specifications, 27
instrumentation front-ends, 18
interconnect buses, 52
isolation, 16
isothermal block, 26

L

latency, 48
    bus, 50
    operating system, 49
    SCSI bus, 53
linearity
    specification, 31
local bus (VXI), 71
    analog signals, 72
    data flow, 75
    digital data path, 74

M

mainframe
    VXI, 82
multi-hit transient recorders, 61
multibuffering, 54
multiple sampling rates, 60
multiplexing, 45

N

noise
    ADC quantization, 38
    pickup, 13
    sources of, 23
    specification, 31
normal mode, 15
Nyquist frequency, 9–10

O

operating system
    latency, 49
    realtime kernel, 56

P

ping-pong buffering, 55
POSIX, 7
processors, dedicated, 56

R

realtime kernel OS, 56
reference voltage, 64
resolution, 45

S

sampling
    data, 9
    theorem, 9
sampling rate
    establishing, 12
    integral sub-multiples, 60
    multiple, 60

scanning ADC, 46
SCSI, 5
SCSI bus, 52
    latency, 53
sensor connections, 13
shielding methods, 18
shunt calibration, 26
sigma-delta converters, 44
signal
    aliasing, 10
signal characteristics, 38
signal conditioning, 25
    bridge, 25
        completion, 26
        excitation, 25
        shunt calibration, 26
    thermocouples, 26
signal to noise
    ADC, 38
simultaneous sampling, 46
single-ended input, 13
software, 67
    large systems, 69
    POSIX, 7
specification
    accuracy, 30
    analog, 27
    bridge completion, 30
    CMRR, 31
    cutoff frequencies, 29
    gain, 29
    gain stability, 30
    input, 27
    linearity, 31
    noise, 31
    shunt calibration, 30
    VXI standard, 81
standards, 5
    CAMAC, 5

GPIB, 5
hardware, 5
    advantages of, 6
    buses, 6
    interconnects, 6
    networks, 6
    SCSI, 5
    software, 6
        POSIX, 7
    VME, 5
    VXI, 5
switched capacitor filters, 37

T

thermocouples, 26
    isothermal block, 26
throughput, 47
time stamp, 60
transient
    data acquisition, 60
    recorders, 60
        multi-hit, 61
trigger lines (VXI), 75

V

VME, 5
VME bus, 51
VXI, 5
    bus, 51
    bus features, 71
    geographic addressing, 85
    local bus, 71
        analog signals, 72
        data flow, 75
        digital data path, 74
    mainframe, 82

INDEX

message-based, 85
module ID lines, 84
module sizes, 82
register-based, 85
slot-0 controller, 83
standard specification, 81
system calibration, 78
system configuration validation, 79
system considerations, 72
system diagnostics, 78
trigger lines, 75
VME relationship, 81

W

wiring practice, 17